Everyday Meteorology

The Thames in flood near Staines

Everyday Meteorology

A. Austin Miller, D.Sc., F.R.G.S.
Late Professor of Geography, University of Reading

and

M. Parry, M.Sc., F.R.Met.S.
Senior Lecturer in Geography, University of Reading

Hutchinson of London

Hutchinson & Co (Publishers) Ltd
3 Fitzroy Square, London W1

London Melbourne Sydney Auckland
Wellington Johannesburg Cape Town
and agencies throughout the world

First published May 1958
Seventh impression July 1972
Second (revised) edition 1975

Set in Monotype Times
Printed in Great Britain by The Anchor Press Ltd
and bound by Wm Brendon & Son Ltd
both of Tiptree, Essex

ISBN 0 09 121910 8

Contents

Plates

Acknowledgements

Acknowledgements are due to the following for permission to reproduce illustrative material.

Plates:

Aerofilms (Frontispiece, 7, 20). B.B.C. (6), Eidg. Institut für Schnee und Lawinenforschung (22), the Headmaster of St. Crispin's School, Wokingham (1), Kemsley Picture Service (21), Institut für Meteorologie und Geophysik der Freien Universität Berlin (2), Murray Hardy (10), F. H. Ludlam (11), C. E. May and Son (5, 14), D. R. Maxted (17), Miss P. Peacock (18), E. Pestell (16), R. A. G. Savigear (19), R. H. Simpson and the American Meteorological Society (15).

Figures:

Donald L. Champion (9.6) and the Director-General of the Meteorological Office and the Controller of Her Majesty's Stationery Office for the use of Crown Copyright material in the Daily Weather Report, Daily Aerological Record and Monthly Weather Summary (4.3, 4.4, 4.5, 4.6, 4.8, 5.2, 5.3, 5.4, 5.5, 5.6, 5.7, 6.2, 6.4, 6.5, 6.7, 6.8, 6.10, 6.11, 6.12, 7.2, 7.6, 9.1, 11.1, 11.3, 11.4, 12.1).

Preface to the Second Edition

A good deal has happened in the field of meteorology since this book was first published. Certain notions are now outmoded, others have become more fashionable: remote sensing has added a new dimension to weather observation: the computer has revolutionized forecasting techniques. Such developments alone justify a restatement of much of what was written fifteen years ago. This second edition is also concerned however with re-emphasizing aspects of the weather picture that have hardly changed. The move towards automation, in observing, in analysing, in forecasting, highly significant though it is, has not eliminated the human element. Human beings, although now spared some former routine labours, still have to interpret the cloud photographs from the satellite cameras, to devise the programmes for the computers and to translate the output of much automated activity into language comprehensible to the community at large.

What has not changed at all is the weather-sensitivity of the man-in-the-street in his various capacities as gardener, exterior decorator, outdoor sportsman, long-distance traveller, holiday-maker or Sunday School outing organizer, let alone his more acute vulnerability as a man-at-work, whether on the farm or the building-site or on the high seas or in the air. No business man in the privacy of his office, no housewife in the seclusion of her own home, no commuter on the train to or from town, not one of us in fact can opt out of our status as 'consumer' vis-à-vis the weather and indeed from some points of view, our susceptibility seems to be greater now than ever it was. No one who has lived through recent winters in this country can be unaware that, when our energy supplies are curtailed, as they may be through industrial disputes or international politics, the difference between muddling through with no more than an occasional inconvenience on the one hand and suffering real hardship and economic chaos on the other may depend very largely on whether the mid-winter months are mild or cold. In New York on the other hand, the lights are likely to go out during summer heat waves when the desire to keep comfortably cool indoors via the air-conditioning plants puts an impossible strain on the power supplies.

The weather is too much a matter of everyday concern to leave entirely to the forecasters. This is not to join in the often unenlightened criticisms directed against these most under-appreciated of our public servants, but rather to suggest that it is our responsibility to get the best out of the forecasts they provide. This requires some understanding on our part of the way the weather works, some familiarity with the terms used by the forecasters and some appreciation of the difficulties of the forecasting problem. The principal aim of this book is to help the reader along these paths.

The text in this second edition has been up-dated in various respects, there has been substantial re-writing and re-drawing of illustrations, but the approach of the book remains unchanged. It attempts a description and explanation of everyday weather situations and events in qualitative terms, which will be of use to the layman who may have forgotten most of his school physics but has not lost his curiosity. The emphasis, as far as the weather of our own latitudes is concerned, is still largely on air masses and fronts and frontal disturbances, regarding these as convenient units of our daily weather experience. The view remains one essentially of weather from the ground, although the upper air receives due mention, of the weather observations anyone can make, although the growing contribution of the satellites is acknowledged, and of the weather forecasts anyone can attempt, although the limitations are not glossed over. The book is the work of geographers with meteorological training, a point that might be thought significant not only because geographers are among those accustomed to taking synoptic viewpoints but perhaps more so because they are very much concerned with the atmosphere as environment, with weather as sometimes a constraint, sometimes a benefit, but always something to be reckoned with by the community in its everyday activities.

There are inevitable changes as regards units of measurement so that parents can keep up with their children. Usually, though not slavishly, two scales are given in the text, temperatures in degrees Celsius or Centigrade (°C) as well as Fahrenheit (°F), distances in metres (m) and kilometres (km) as well as feet and miles, speeds in metres per second (m/sec) as well as miles per hour. Weather maps are now drawn broadly in the style of the Daily Weather Report maps. Although this is not intended as a course book, much of the illustrative material of this kind, derived as it is from actual situations, lends itself to adaptation into simple exercises in weather-map analysis and appreciation, which teachers may find directly useful.

M.P. 1975.

1 The Raw Material of Weather Studies

Observations for All

Measurements and observations of the state of the atmosphere are the raw material of the study of weather. Appraisal of the atmosphere dates back to the very beginnings of man's consciousness but, as a physical science, meteorology began in the seventeenth century with the invention of instruments such as the thermometer and the barometer. To-day there is a vast range of meteorological sensing equipment, some of it of a complexity baffling to the layman, but it remains true that the atmosphere is a free laboratory, open to us all, in which we can measure a great deal with sufficient accuracy, provided we are prepared to take a little trouble.

It is very easy to hang up a thermometer out of doors in the sunshine, and to delude ourselves that we are making a meaningful assessment of the warmth of the day. The fact is that different substances exposed to the sun's radiant heat warm up to differing extents that depend on their physical properties. The air is warmed very little in this way but a thermometer in the sun warms up considerably and what we read is a temperature much higher than that of the air. If we require our thermometer to register a faithful air temperature, we must ensure that it is shaded from direct solar radiation. At the same time, it must not be solidly enclosed: it needs to be well ventilated, that is, cut off as little as possible from the free-moving air. With these precautions taken, we have some approach to a representative air temperature.

A time-hallowed answer to this problem has been to hang the thermometer on a north-facing wall, but this does not meet the requirements of good ventilation or of complete shade (in the summer months), quite apart from the possibility of heat seepage through the wall. The standard answer is to house the thermometer in a box of approved design with louvred sides to ensure reasonable ventilation (Stevenson

Screen) but this is relatively expensive. A very adequate answer for the interested amateur is the so-called whirling hygrometer (Fig. 1.1) which has the advantages of being cheap (about £5), portable and providing information about humidity as a bonus. It consists of an ordinary ('dry-bulb') thermometer and a 'wet-bulb' thermometer (see p. 21) mounted in a frame very much like a football supporter's rattle

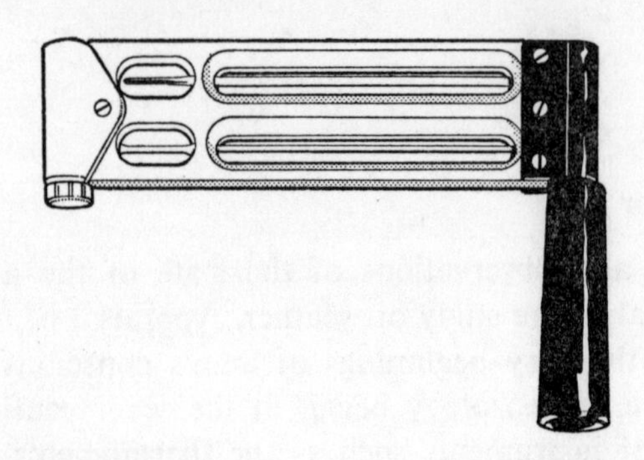

1.1 A whirling hygrometer.

and it is used in the same sort of way. Rapid whirling of the instrument provides the necessary ventilation and this operation is repeated until the temperature remains the same at two consecutive readings. In direct sunlight the body is used to shield the instrument: on a cloudy day it matters little. Properly used, the 'whirler' will register the environmental temperature fairly faithfully, but it is still up to the user to select the most representative recording 'station', which usually means the middle of the garden or some nearby open space.

The measurement of rainfall is equally feasible and for a very small outlay, since a rain-gauge is simply a receptacle exposed so as to catch falling rain. Conveniently this takes the form of a funnel with its spout leading into a glass measuring jar or bottle and the cheapest commercially available gauges (costing under £5) are no more than this. There is an obvious relationship between the size of the mouth of the funnel, the width of the collecting vessel and its calibration for measurement purposes, which, for a do-it-yourself device, will necessitate a little arithmetic, but it is worth remembering that the most commonly used standard gauge in the country has a 5 inch (127 mm) diameter aperture, for which measuring jars carefully calibrated in millimetres (or parts of an inch) are readily available for the price of £1 or so. As with temperature readings the information provided by a rain-gauge, home-made or otherwise, will depend very much on its location. Close proximity to buildings or trees will rob the gauge of some of its rightful catch. The whole problem of rain-gauge exposure is a tricky one, and we examine it more fully later on: for the moment, suffice it to say that a position in an open part of the garden lawn, with the funnel fixed to a post with its mouth about a foot off the ground, is likely to give the best results.

We now turn our attention to wind. There are instruments for measuring wind speed (anemometers) but these are expensive and do not give full value unless the exposure is suitable. A quite useful estimate of wind force may be made in terms of the Beaufort Scale (based on a scheme originally devised early last century by a British admiral for use at sea), which is a series of descriptions of the way common outdoor features like smoke, wind vanes, leaves and branches of trees, respond to different wind conditions. The Beaufort signs for land are reproduced in Table 1: the most usual categories are very quickly learned. Wind direction is commonly registered by wind vanes (which can easily be home-made) but these are rarely mounted high enough or freely enough to escape the considerable distorting influence of obstacles on air flow: a light flag or a home-made wind-sock on a suitably placed pole or smoke from tall chimney stacks, which will show the direction even of very light winds, will give better results (except that, fortunately from all other standpoints, few tall chimney stacks smoke these days).

Admiral Beaufort also devised a series of initial letters which form a very convenient short-hand for the description of certain visual aspects of weather. A selection of these is given in Table 2. There remains the identification of cloud types, satisfying in itself and invaluable for the appreciation of the weather processes of the moment and the portents for the future. Descriptions of cloud types are given in later pages. It is also useful to make a rough estimate of the amount of cloud, in 'oktas' (eighths) of sky covered.

An important weather element that cannot be assessed by the senses or measured by improvised means is atmospheric pressure. However, the 'hall barometer' is perhaps the most ubiquitous of meteorological instruments and, provided we ignore the legends 'Rain', 'Change', 'Fair', etc, that usually appear on the dial, we can make good use of it if we have one. This is a so-called aneroid type, consisting basically of a thin-walled metal biscuit-like capsule partly exhausted of air: one side of the capsule is fixed, the other is more or less depressed in response to changing atmospheric pressure. The tiny movements so induced are magnified many times by a system of levers and arranged

Table 1 Beaufort Scale of Wind Speed

Force	Description	Effects on Land	Equivalent Speed M.p.h.	Equivalent Speed m/sec
0	Calm	No wind: smoke rises vertically	<1	0–0·2
1	Light air	No noticeable wind: vanes remain still, but smoke drifts	1–3	0·3–1·5
2	Light breeze	Wind felt on face: vanes move, leaves rustle	4–7	1·6–3·3
3	Gentle breeze	Hair, clothing disturbed: light flag extended, leaves and small twigs move constantly	8–12	3·4–5·4
4	Moderate breeze	Hair disarranged: raises dust and loose paper, moves small branches	13–18	5·5–7·9
5	Fresh breeze	Disagreeable wind, force felt on body: small trees in leaf begin to sway	19–24	8·0–10·7
6	Strong breeze	Difficult to walk steadily, umbrellas hard to control: large branches move, telegraph wires whistle	25–31	10·8–13·8
7	Near gale	Some resistance to walkers: whole trees move	32–38	13·9–17·1
8	Gale	Impedes progress: breaks twigs off trees	39–46	17·2–20·7
9	Strong gale	People blown over by gusts: removes roof slates and chimney pots	47–54	20·8–24·4
10	Storm	Rare inland: trees uprooted, much structural damage	55–63	24·5–28·4
11	Violent storm	Rare: widespread damage	64–72	28·5–32·6
12	Hurricane	Rare	>73	>32·7

to move a pointer over a calibrated circular scale. Alone among meteorological instruments, the barometer performs its task indoors and the hall is usually as good a place as any other for it.

Table 2. Weather Description by Beaufort Letters

Appearance of Sky	b blue sky, bc partly clouded, c cloudy, o uniformly overcast
Precipitation	d drizzle, r rain, s snow, rs sleet, p shower, h hail
Electrical Phenomena	l distant lightning, t distant thunder, tl thunderstorm
Atmospheric Obscurity	m mist, z haze, f fog, fe wet fog, F thick fog
Ground Phenomena	w dew, x hoar-frost

Note: Beaufort Letters are combined as follows:

Capital letter denotes intense e.g. R	= heavy rain
Letter repeated denotes continuous e.g. dd	= continuous drizzle
Suffix subscript o denotes slight e.g. s_o	= light snow
Prefix i denotes intermittent e.g. ir	= intermittent moderate rain
Prefix p denotes shower of ... e.g. ps	= snow shower

Prefix j denotes weather near but not at the station.
Solidus / divides present weather from weather in preceding hour e.g. r/tl = rain now but thunderstorm in last hour.

To sum up, it is well within the reach of the interested layman, at small expense, to record air temperature, humidity and pressure and it costs exactly nothing to observe wind direction and speed, cloud and weather. This constitutes a very fair sample of the atmospheric environment. These observations will provide a constant source of interest, speculation and conversation with the neighbours. Of course, the amateur weather-watcher may go further: more than one, blessed with time, money and suitable garden space, runs a fine station, obeying all official requirements. Most of the nearly 7000 rain-gauges in this country, of which monthly returns are sent to the British Rainfall organization (a branch of the Meteorological Office), are maintained by amateurs. But for the amateur with such aspirations, there is, of course, much more to learn.

A School Weather Station

This section is intended to assist such a person and, particularly, the teacher who is considering the establishment of a school weather station which would both serve admirable educational purposes and yield valuable local records. There is no reason why he should not aim high: many schools run surprisingly well equipped weather stations (Plate 1) and some even report to the Meteorological Office. But if this is the aim, the station must be properly sited and maintained and the observations carried out with complete regularity and reliability. What follows can be no more than an introduction: the Meteorological Office publication *The Observer's Handbook* is the essential and indispensable reference book.

The first of a number of exacting requirements is a suitable site, that is to say, an open level lawn (close cropped grass is a standard meteorological surface for land observations) which allows a correct exposure for the outdoor instruments. 'Correct' here means representative of the area, which in turn means freedom from highly localized peculiarities, such as close proximity to the excessive shelter of trees or the undue warmth of sun-lit walls or the stagnant conditions of a small walled-in garden or interior courtyard. The accepted convention is that, given an obstruction (wall, building, tree etc) of height x, then both the thermometer screen and the rain-gauge should be at a distance of at least $2x$ from it. (Some authorities favour a $4x$ rule for the rain-gauge.) All too often, such a site turns out to be available only in the middle of the school playing fields and from this misfortune complete recovery is not possible. The teacher will then have to secure the best alternative site, in the knowledge that the records yielded will still have great educational value but will not be comparable with others from properly exposed stations.

Proceeding optimistically, we now turn to the various instruments in more detail. The approved thermometer housing (Stevenson Screen) is a wooden box with louvred sides, double roof and a floor consisting of overlapping boards (not in contact): all this allows maximum ventilation. The whole is painted white and the hinged door is on the north-facing side (in the northern hemisphere): this minimizes the effect of solar radiation. The screen is mounted on a stand of such a height that the thermometer bulbs are a standard 122 cm (4 ft) above the ground: this convention recognizes the existence of often strong temperature gradients in the lowest air layers and thus ensures comparability.

A commercially available Stevenson Screen costs most of £30. At least one well known manufacturer sells a 'make-it-yourself' construction kit, for half that figure, that assembles to the same product. An even bigger challenge to the woodwork class is offered by the pamphlet, M.O. 670. *Instructions for making the Stevenson type of Thermometer Screen*, available from Her Majesty's Stationery Office. Thermometers must be purchased and these usually cost from £3 to £6 depending on type. Probably, those with the stem mounted on a wooden or plastic back-board and the bulb protected by a metal guard, are likely to have the longest expectation of life.

The screen houses four thermometers (Fig. 1.2). Two are of ordinary

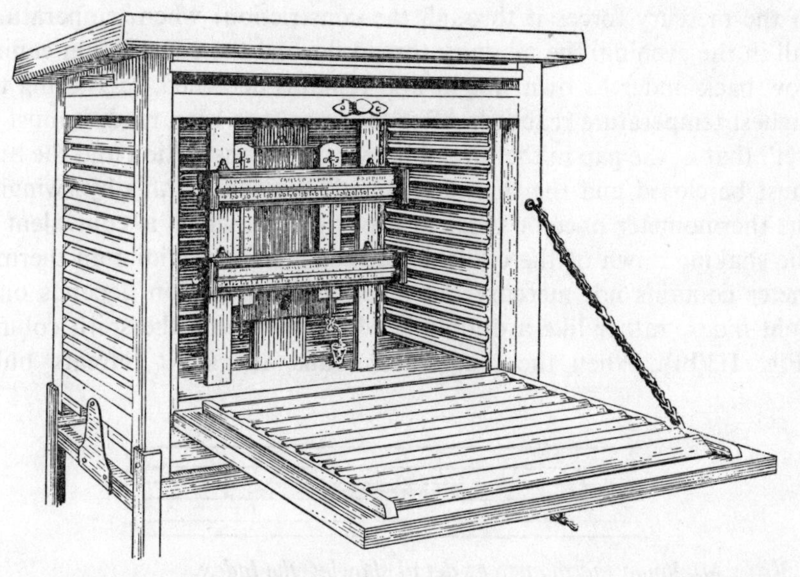

1.2 Stevenson Screen, open to show the arrangement of the thermometers.

mercury type, the 'dry-bulb' and 'wet-bulb' thermometers, the latter having its bulb encased in a muslin bag kept moist by a cotton wick trailing into a container of distilled water: this is the equivalent of the 'wet bulb' in the whirling hygrometer, which has a wick in the form of a sleeve dipping into a plastic water reservoir. The two thermometers give respectively, for the time of observation, the 'dry-bulb' (ordinary air) temperature and the 'wet-bulb' temperature (see p. 67), from which may be derived an expression of the humidity of the air. The muslin and wick quickly become dirty, especially in towns and may require changing perhaps once a week.

The other two thermometers are self-registering: they provide maximum and minimum temperatures over a given period. The maximum thermometer (of which the well known clinical thermometer is a specialized type) has a constriction in the bore near the bulb end (Fig. 1.3(a)). During the morning rise of temperature, the expansion

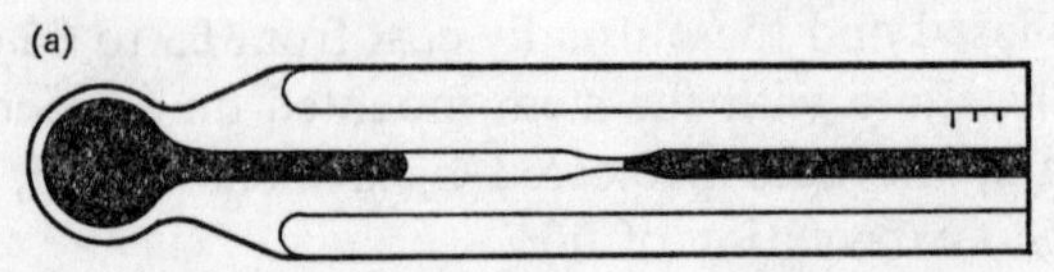

1.3(a) Maximum thermometer: detail showing the constriction in the bore.

of the mercury forces it through the constriction: when temperatures fall in the evening, the mercury thread beyond the constriction cannot flow back under its own weight and remains detached, registering the highest temperature reached. After the 'max' has been read, it must be 'set', that is, the gap in the thread between the constriction and the bulb must be closed and the thread reunited. We achieve this by swinging the thermometer once or twice at arm's length, which is equivalent to the shaking down of the clinical thermometer. The minimum thermometer contains not mercury but alcohol: its operation depends on a light index, rather like a double headed pin, within the spirit column (Fig. 1.3(b)). When the temperature falls, the spirit retreats bulb-

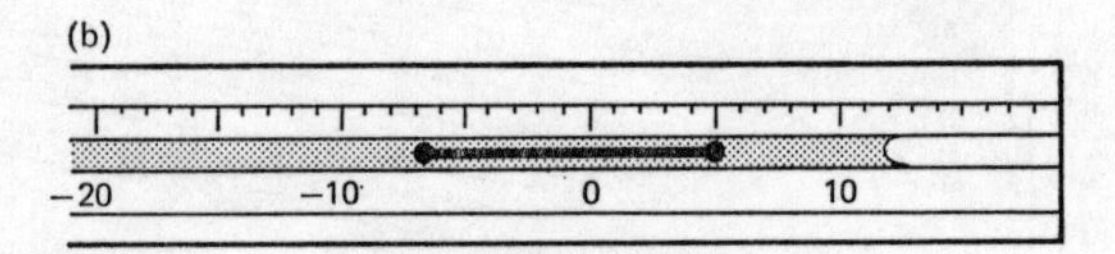

1.3(b) Minimum thermometer: detail showing the index.

wards and drags the index with it by a surface tension effect: when the temperature rises, the spirit advances, ignoring the index which is left to register the lowest temperature, normally reached during the night. What we read at the morning observation is not the end of the spirit column (which gives the present temperature) but the adjacent end of the index and, after reading, we reset the 'min' by tilting it so that the index slides back into contact with the end of the liquid.

A minimum thermometer of sheathed pattern (that is, with the graduated stem enclosed within an outer protective glass sheath) can serve to measure 'grass minimum' temperatures. The thermometer is exposed at night on the lawn, supported by two pegs or twigs, so that the bulb is in contact with the grass tips: it is the source of 'ground frost' data. At more specialized stations, a similar thermometer exposed on a bare patch acts as 'bare ground minimum' and – an interesting recent development – one placed on a concrete slab gives 'concrete minimum temperatures', useful in connection with road surface studies and problems of urban climate.

A description of the construction and emplacement of the British Meteorological Office design rain-gauge will illustrate some of the problems of rainfall measurement. As Fig. 1.4 shows, the metal (copper

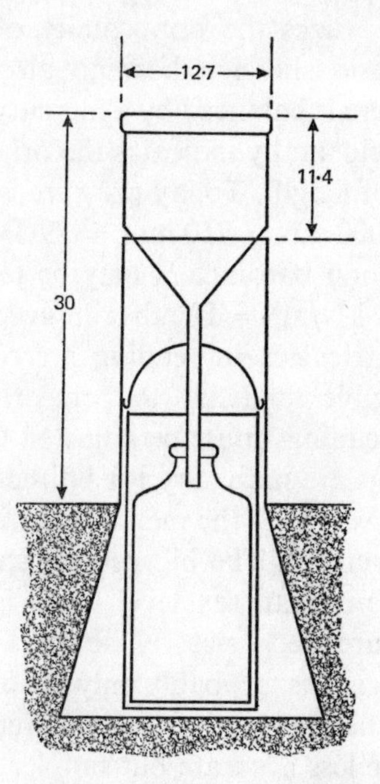

1.4 Rain-gauge: cross-section diagram (dimensions in cm).

or galvanized-iron) funnel which has a deep rim to prevent losses by splashing and a small diameter spout to minimize evaporation, fits into and is supported by a cylindrical can which is partly sunk into the ground. The collecting bottle stands in an inner vessel which takes the excess in the rare case of the bottle overflowing. To measure the catch we pour into a glass measuring cylinder of scale appropriate to the diameter of the funnel mouth.

The standard exposure of the rain-gauge in this country has the funnel rim exactly 30 cm (12 in) above ground. This height is a compromise between the demands of different considerations. Nearest the ground the wind strength is at its least, rain tends to fall most nearly vertically and the maximum catch is recorded. Higher up, the wind is stronger and subject to eddies and swirls, blowing the rain across the mouth of the gauge and thus reducing the catch: it is for this reason that roof-top sites are not recommended. At the other extreme, if the gauge were completely buried, with its rim flush with the ground, rain could well splash in from outside. Interest, in fact, is growing in a 'ground gauge' which does have its mouth at ground level but stands on the floor of a pit with a non-splash 'roof' of egg-crate character.

Most schools will possess a mercury barometer of some sort and every science teacher seizes the opportunity of repeating Torricelli's original experiment and showing how an air column extending the depth of the atmosphere is balanced by a mercury column some 760 mm or 30 in high. This incidentally indicates the origin of the measurement of pressure in terms of length. Today pressure is expressed as a unit of force, the millibar (1000 mb = 750 mm = 29·53 in roughly), although we are warned that soon barometers may be re-scaled in kilonewtons per square metre ($1 \text{ kN/m}^2 = 10$ mb) in accordance with the S.I. system. There are intricacies in reading mercury barometers, partly due to their incorrigible tendency to behave as thermometers. For accuracy, the crude readings must be adjusted to a standard temperature. Other corrections are necessary for latitude, with which the force of gravity (hence the weight of the mercury) varies and for the 'individual error' of the instrument. The biggest adjustment is for altitude, if, as is usual, a common mean sea level value is required: in the first 1000 m or so, pressure decreases by about 1 mb in every 10 m of ascent. For all these reasons, probably only the older pupils will manage the complexities of the standard pattern mercury barometer and for younger observers the less accurate but simpler aneroid will do.

For the rest, the eye observations and estimations dealt with in the preceding section will again be relevant. The school might possibly acquire an anemometer of sorts but it is worth remembering that the use of the Beaufort Scale gives better training in observation than the reading of a pointer on a dial. What has been described so far would certainly constitute a very useful school weather station. There remains the problem of organization, for which every teacher will have his own solutions. Observation time should not vary and, if it can be 0900 hr GMT, this will be in line with other voluntary recording stations (though some observe more than once daily). Of course we have by no means yet exhausted the range of observational resources: to learn more we need to visit an official 'met.' station.

1. The weather station at St. Crispin's School, Wokingham. Much of the equipment, including the scale model of the 'Skylon' carrying the anemometer and wind-vane, was made by the pupils.

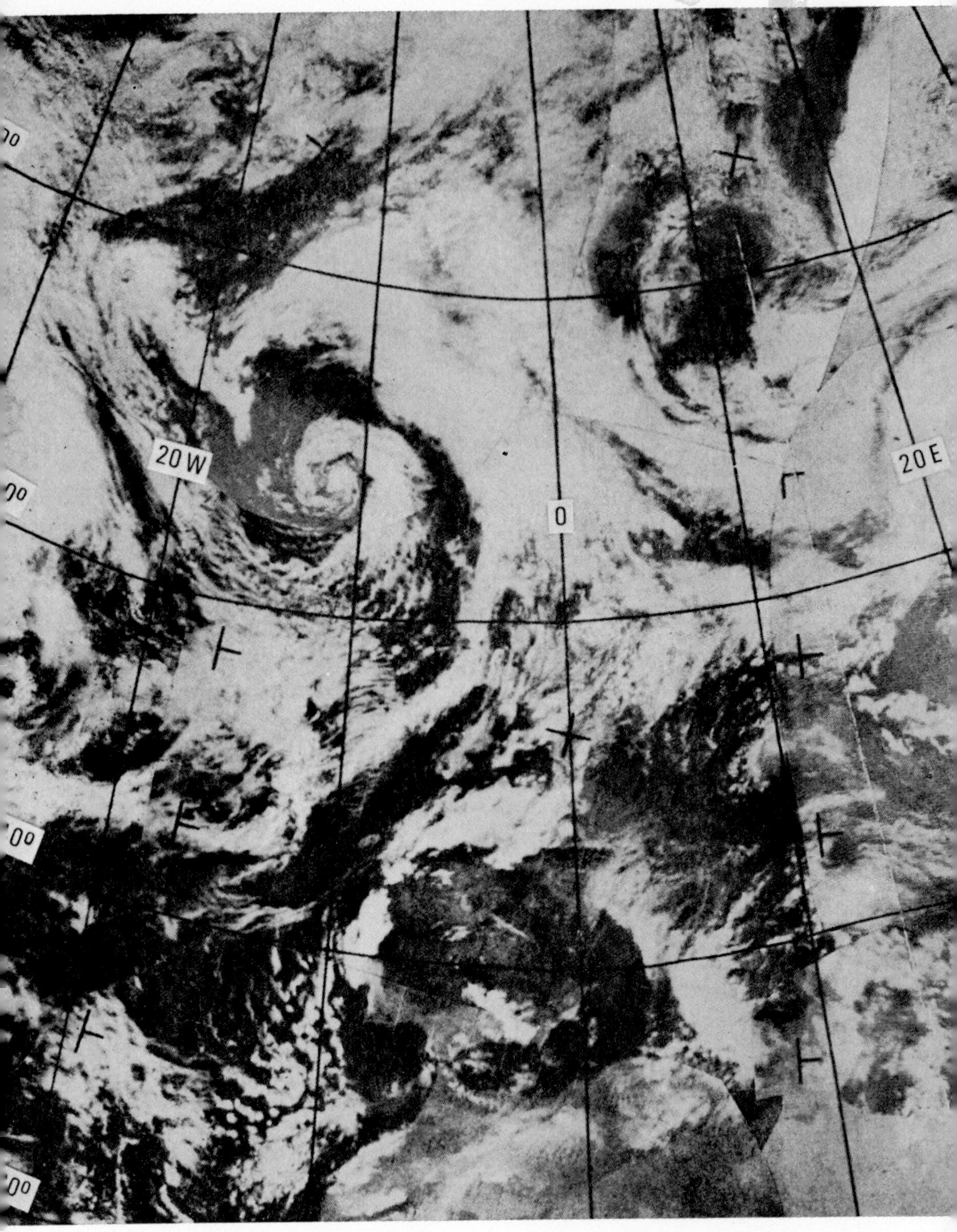

2. Part of a satellite cloud photograph mosaic: 26th September 1971. Skies are clear over southern Spain and Morocco but Britain is hidden by cloud. A frontal depression is centred west of Ireland.

3. Polar maritime weather: an afternoon sky laden with cumulus clouds.

4. Polar maritime weather: the sun breaks through heavy clouds after a shower.

5. Polar continental weather: Whiteknights Lake, Reading, early 1947.

Official Weather Stations

The organization of weather observing in this country is in the hands of the Meteorological Office, which is part of the Air Force Department of the Ministry of Defence. Stations that report regularly to the Met. Office vary from very fully equipped and professionally staffed synoptic stations (of which there are over 100) providing in the main, hourly information for forecasting purposes and located usually on airfields, to more than 500 climatological stations, maintained by observatories, research institutions, local authorities, universities and some schools, where information, less urgently required, is sent as monthly returns. Some of these voluntarily maintained stations are very well equipped and carry out specialized observations for research or other needs. Nearly 100 'agro-meteorological' stations have special responsibilities particularly with regard to measuring soil temperatures. Resort stations are specially concerned to report on their maximum temperatures and sunshine hours.

These stations will generally possess much more than the basic equipment already described. Most synoptic stations have a well exposed anemometer and wind-vane, at the internationally accepted standard height of 10 m (33 ft) above open level ground (or higher where obstructions must be taken into account), to give accurate measurements of wind speed and direction. A standard type of anemometer consists of three metal cups attached to arms mounted on a vertical spindle (Fig. 1.5): the cups rotate in the wind, at a rate proportional to its speed, and the motion activates either a counter mechanism (cup counter type) or a small electric generator (cup generator type): the latter shows wind speed on the indicator dial of a voltmeter

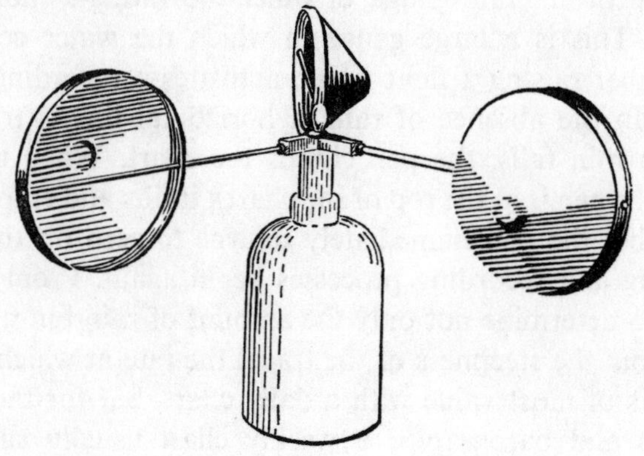

1.5 Cup anemometer.

in disguise, i.e. one calibrated in knots, or miles (or kilometres) per hour or metres per second. The wind vane too can be made to indicate, with the pointer traversing a scale calibrated in degrees from north.

Much of the additional equipment at these stations is of the continuous recording type. In all these instruments the recording element, whatever it may be, controls the movement of a pen which marks a continuous ink trace on a chart wrapped round a drum rotated by clockwork. For example, in the thermograph, which gives a continuous temperature record, the thermal element is a coil formed of two strips, welded face to face, of different metals which expand at differing rates when heated. This bi-metallic coil responds to temperature changes by uncoiling or coiling tighter and this mechanical action is harnessed to activate the pen. The hygrograph, which gives a continuous humidity trace, makes use of a long-recognized property of human hair: the length of a hair, if cleaned of grease, varies with changing humidity. In this instrument the change in the length of a specially prepared bundle of hair operates the pen through an ingenious mechanism. Thermograph and hygrograph are usually accommodated, with the four thermometers, in double-size Stevenson Screens. In the barograph, several aneroid elements are linked to provide the mechanical movement responsible for the pressure trace, which is especially valuable since pressure changes over 3-hour periods form an important item in the weather messages sent by synoptic stations. The output from electrically recording anemometers and wind vanes can also be adapted to give wind speed and direction traces: to watch this recorder in action is to be convinced of the restless nature of the wind. Some examples of such traces are shown in Fig. 6.10.

Also shown in that diagram is the kind of trace yielded by the autographic or recording rain-gauge, of which the 'tilting-siphon' type is the most usual. This is a large gauge in which the water collected falls into a chamber raising a float which activates a recording pen in the usual way. In the absence of rain, a horizontal line is traced on the chart: when rain falls, the pen climbs the chart. When the chamber is full (and the pen is at the top of the chart), it tilts and empties through a siphon tube: the pen immediately returns to zero on the chart and the collecting and recording processes begin again. From the chart it is possible to determine not only the amount of rain but also its duration and (from the steepness of the trace) the rate at which it fell. This instrument is of most value with a daily chart, but for thermographs, hygrographs and barographs, a weekly chart usually suffices, while the wind recorder operates on a monthly roll chart.

A permanent record of a different sort, the duration of bright sunshine, is provided by the Campbell-Stokes recorder (Fig. 1.6). The principle is simple enough: a glass sphere, acting as a lens, focusses the sun's rays on to a sensitized card held in a metal half-bowl with which the sphere is concentric. With the apparent movement of the

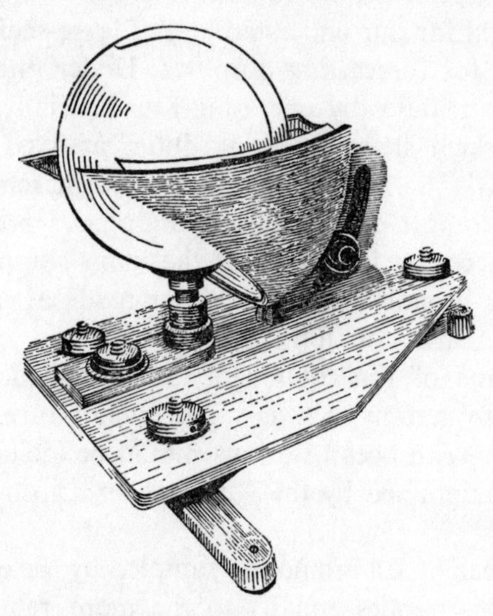

1.6 Campbell-Stokes sunshine recorder.

sun across the sky, a brown trace is scorched on the blue card and, as the latter is calibrated in hours, the duration of sunshine, whether from a continuous trace or one interrupted by cloudy periods, may be read off. The card is replaced daily, even if, as may well happen in this country, it shows no burn at all. The recorder requires the most open situation available and a roof-top is often best.

A full range of supplementary eye observations and estimations is made, particularly at synoptic stations, since such details may be of great significance to aviation. These include cloud type, amount and height of base and also visibility, which is estimated by noting which of a pre-determined series of landmarks (varying from nearby buildings to distant spires or hill-tops) at known distances from the station remains recognizably visible from some suitable viewpoint. Incidentally, there is no reason why the amateur or the school observing team should not try visibility estimation. At night, observers rely on the visibility of fixed lights at known distances.

Weather Stations at Sea

Now taking a global view, we may note that there are more than 10 000 synoptic reporting stations over the land area of the world, though these tend to be concentrated in the more developed regions. However, over 70% of the world surface is sea and information from these areas is vital for our understanding of large-scale weather events and particularly for forecasting purposes. Under international agreement, Britain contributes by operating four specially equipped ocean weather ships which share reporting duties at fixed positions in the eastern North Atlantic. Worldwide, over 5000 merchant ships voluntarily make and report observations using agreed procedures, though this information comes largely from the main shipping routes. Ship observations are broadly similar to those made at land stations but certain problems arise on a floating station and modifications are necessary in some of the equipment and techniques: on the other hand, special information such as water temperature and state of sea can be provided. With ocean stations should be included a number of polar stations maintained by the Russians on drifting ice floes in the Arctic.

Since large areas of sea are poorly sampled by the existing observing network (and this applies equally to the more remote land areas), there is considerable interest in the development of automatic weather stations. These are feasible because temperature, rainfall, wind speed and direction can all be sensed so as to produce an electric signal which can be recorded on a moving chart and the device can be battery operated. A number of these unmanned stations exist already in several land locations, operating experimentally, and a few similar devices have been mounted on navigation buoys. The full advantages of this extension of automation await the perfection of methods of retrieving the information in the most effective manner.

Weather Information from the Upper Air

Weather is a three-dimensional phenomenon and the answers to many problems lie in conditions high above the earth's surface. This was well understood by the pioneers who first probed the upper air in hazardous balloon ascents in the 1850s and 1860s. Later, kites and captive balloons, carrying self-recording instruments, were sent up, sometimes at the end of 5 or 6 miles of wire. Between the World Wars, spiralling ascents by specially equipped aircraft became the main source of upper-air data: the observer recorded pressure, temperature and humidity at various heights and could also report on conditions of cloud, icing and weather. Today, however, aircraft flights are used mainly for specialized research purposes and the routine task of measuring upper-air conditions is left to radio-sounding methods.

The British radio-sonde, in use since 1939, consists of a balloon (some 2 m in diameter) which carries up a multiple recorder – an aneroid capsule for pressure, a bi-metallic coil for temperature and a strip of goldbeater's skin sensitive, like human hair, to humidity – and also a battery-powered radio transmitter which sends the information in the form of an audio-frequency note to the ground receiving station. Each recording unit is connected in turn to the transmitter by means of a switch operated by a small windmill. The result of changes in pressure, temperature or humidity is to vary the pitch of the note transmitted. At the ground station, the observer tunes his receiver to the signal and translates from the frequency of the note into familiar pressure, temperature or humidity values: each element 'comes round' about once a minute. At a height varying from 15 to 30 km (say 10 to 18 miles), the much distended balloon bursts and the equipment returns gently to earth by means of an attached parachute.

The same balloon can provide the means of measuring upper winds (replacing older methods of 'following' a pilot balloon of assumed rate of ascent through a specially designed theodolite): now the balloon carries a wire-mesh reflector which is tracked by radar. There are eight aerological stations in the British Isles and the ocean weather ships carry out the same programme.

A number of radio and radar techniques are employed in the many-fronted attack on problems of meteorological observing. For example, lightning activity is detectable by radio: the flashes generate radio transmissions which are picked up as 'atmospherics' on our domestic receivers and equally register on a cathode-ray oscillograph. The British thunderstorm location service 'fixes' these flashes up to about 2500 km (1500 miles) away by means of four direction-finding stations. Radar is used to locate rain-bearing systems of convective storm type, since large rain-drops give echoes which show up on the radar screen. Two types of equipment are in use: the Plan Position Indicator gives the bearing and distance of the 'radar cloud' from the station, with a range of about 160 km (100 miles) utilizing a radar beam scanning horizontally: the Range Height Indicator scans vertically, giving the base and top of echoes. These devices are invaluable for short-term forecasting in suitable situations.

Above the limit of the radio-sonde balloon, rockets, with meteorological sensing equipment replacing the warheads which were their original raison d'être, extend the opportunities of probing vertically to above 100 km (say 60 miles), mainly as part of research programmes. The other function of the rocket, of course, is to launch the space satellites which from 1960 have added a new dimension to weather observations.

World Weather Watch

These words constitute the title, usually abbreviated to WWW, of an ambitious international programme for a global atmospheric survey, recognizing the limitations of the surface observing network and the inadequate charting of the upper air by the 1000 or so existing aerological stations: it involves the measurement, storage and retrieval of world-wide meteorological data. At the same time the words very well represent the possibilities opened up by ever-improving satellite technology. By now we are all familiar with the fascinating cloud pictures taken from the weather satellites orbiting some hundreds of miles above the earth's surface: such photographs, which strikingly confirm and often amplify the conventionally produced weather maps, were published weekly in at least one national newspaper and are beginning to figure in the television weather forecasts. The satellites are 'instructed' from ground stations, where the transmitted pictures are rebuilt and photographed (Plate 11): when the satellite is out of range of the ground station the information is stored on magnetic tape for playback later.

Cloud photographs (which require careful interpretation) are only the most obvious output of the weather satellite systems. Radiation both from the sun and the earth is measured and infra-red scanning can yield a surprising amount of information about cloud conditions (by night as well as by day) including, via cloud-top temperatures, an estimation of their height. Very promising work is in progress on the construction of vertical temperature profiles (otherwise the function of the radio-sonde ascents) by making use of the absorption characteristics of carbon dioxide in the atmosphere and selecting radiation (the intensity of which depends on temperature) from different heights. There are similar possibilities with regard to water vapour profiles. Furthermore, satellites can be equipped to locate and track suitable signals and they can also be made to 'question' unmanned stations like the earlier-mentioned weather buoys at sea and retrieve information from them.

The observational facilities created by WWW will be utilized for the Global Atmosphere Research Programme (GARP), which, beginning with a tropical investigation in 1974, aims at a fuller understanding of the structure, behaviour and predictability of the atmosphere. In addition to conventional surface and upper-air data and automatic weather stations and remote sensing from satellites, GARP will make use of balloons designed to drift with the air flow at selected pressure levels, carrying sensors which can be 'interrogated' by satellites. Some such GHOST (Global Horizontal Sounding Technique) balloons have already been released experimentally and the results indicate that at lower stratospheric levels (see p. 51) a balloon may last 6 months or even a year but, in the lower troposphere, icing may reduce the life to only a week or two.

Impressive though these projects are and remote as they are from the simple observations outlined at the beginning of the chapter, it should not be thought that the complete answers they seek are necessarily close at hand, or that the day of the human observer is done. Certainly, these programmes are expensive but justifiable in view of the economic benefits of an eventually improved forecasting service. Certainly, they require international organization and collaboration on a massive scale. Organization – and that of a conventional meteorological service is complex enough – will engage our attention in the next chapter.

2 The Processing of Weather Information

Between the initial gathering of the raw weather data and its eventual use as information by a wide range of consumers lies a series of intermediate steps which involve a high degree of organization. Little more will be said concerning reports intended primarily for climatological or research purposes: these are directed into somewhat quieter backwaters of the meteorological organization. What follows in this chapter is mainly concerned with the swift sequence of events from observation to forecast and, while the details given refer particularly to the British meteorological service, the arrangements are essentially similar in most national organizations.

Functions of a State Meteorological Service

The Meteorological Office is charged with the tasks of providing meteorological services for a great variety of users including the armed forces, civil aviation, merchant and fishing fleets, government and local authorities, the Press and the general public, of organizing meteorological observations in the United Kingdom, of collecting, distributing and publishing worldwide weather information and of maintaining a full research programme. At all synoptic and auxiliary reporting stations the observations described in the previous chapter are carried out at standardized times, using standardized equipment and following standardized procedures. The reports are entered in a standard form of observation register which serves as a permanent record. However, for the urgent needs of forecasting this information must be made available as quickly as possible wherever it is received. A rapid system of communications is thus an essential feature of an efficient meteorological service.

No method of rapid exchange of weather messages existed before the advent of the electric telegraph: the first telegraphed weather reports were published in the *Daily News* in 1848. Line telegraphy later gave way to radio-telegraphy, linking distant countries and ships at sea. Within the British Isles, as in many other countries, the main internal means of communication is now by (line) teleprinter. Over 100 British stations are currently linked by the teleprinter network and lines also extend to neighbouring European countries. The same teleprinter signals are freely available to all by radio transmission. For all such methods of data exchange, the weather reports are converted into numerical coded messages. The newest method in telecommunications is photo-telegraphy or facsimile transmission (FAX), an ingenious process by which pictorial information of any kind is 'broadcast' by one station and 'received' at others: FAX transmission can be either by land-line or radio.

In these various ways information is received by the forecaster either in pre-digested form (by FAX) or in coded messages that must be represented in a format allowing an analysis of the current weather situation. There are standardized procedures for the plotting of surface data on prepared outline maps: such a map when completed provides a general view (or synopsis) of the weather over a wide area and is in fact called a *synoptic chart*. The upper-air data similarly find their way on to aerological charts and diagrams of various kinds. Beyond this stage, the organization, having provided the available evidence, recedes somewhat into the background and, to an extent depending on the status of the station, the individual forecaster comes into the centre of the picture. It is his task to appraise and analyse the material, and, as a climax to all this activity, to prepare the required forecasts.

No national meteorological organization can be self-sufficient and it is not surprising that in normal times few services are as effectively internationalized as the exchange of weather information. Co-operation in this field is more than a century old: it was in September 1873 that an international congress met in Vienna, from which sprang the non-governmental International Meteorological Organization (IMO). Notwithstanding its unofficial status, IMO succeeded in building up an international synoptic system, encouraging uniformity in observing, reporting and plotting procedures and co-ordinating arrangements for the exchange of information. In 1951, IMO was transformed into WMO (World Meteorological Organization), now having the status of a specialized agency of the United Nations, with responsibility for internationalizing organized meteorological activity.

At the 'Met. Office'

The British meteorological service is organized in a hierarchy of units and it will be useful to view the structure from bottom and top respectively, i.e. by describing activities first at a subsidiary or satellite meteorological office and then at the Central Forecasting Office. The typical 'met. office' combines the functions of synoptic reporting station and forecasting unit and is usually located on an airfield, RAF or civil. If we visit such an office our first visual impression is likely to be of charts, which cover the bench surfaces and most of the wall space: at the same time we become aware of the constant chatter of the teleprinter (housed usually in an adjoining room). The office is manned normally by a forecaster and one or more assistants, much depending on its status and responsibilities.

Concentrating our attention first on the assistant, we note that he becomes galvanized into activity at approximately ten minutes before the hour: this is the time for the hourly observation or 'ob'. The visit to the instruments, outdoor and indoor, and the expert appraisal of the sky are carried out at great speed. Still before the hour has struck, the 'ob' is duly recorded in the register and also condensed into a coded message of the form shown (with translation) opposite.

At a signal the 'ob' is transmitted on the teleprinter. The assistant types the message on the keyboard and we watch the five-figure groups appearing on the teleprinter paper. They are also appearing simultaneously on a teleprinter at the parent station which controls this particular subsidiary and a number of others and acts as a regional centre for the collection and distribution of reports. If we continue watching we observe the teleprinter chattering on, as the messages from other stations in the group are duly sent and acknowledged.

Every other parent office in the network is at this moment similarly receiving the 'obs' from its respective satellite stations. Once collected, the messages are passed from the regional centres to the national centre, that is, the Central Forecasting Office at Bracknell. This task is completed by about ten minutes past the hour. At that moment, every station on the teleprinter network is switched via its parent office to Bracknell and the entire series of now classified messages is relayed directly to all concerned. After about half-an-hour of British and near-European data, there follow other foreign reports, ships' observations, upper-air data and sundry less regular communications. At five minutes before the next hour the transmission ceases: shortly the cycle will begin anew.

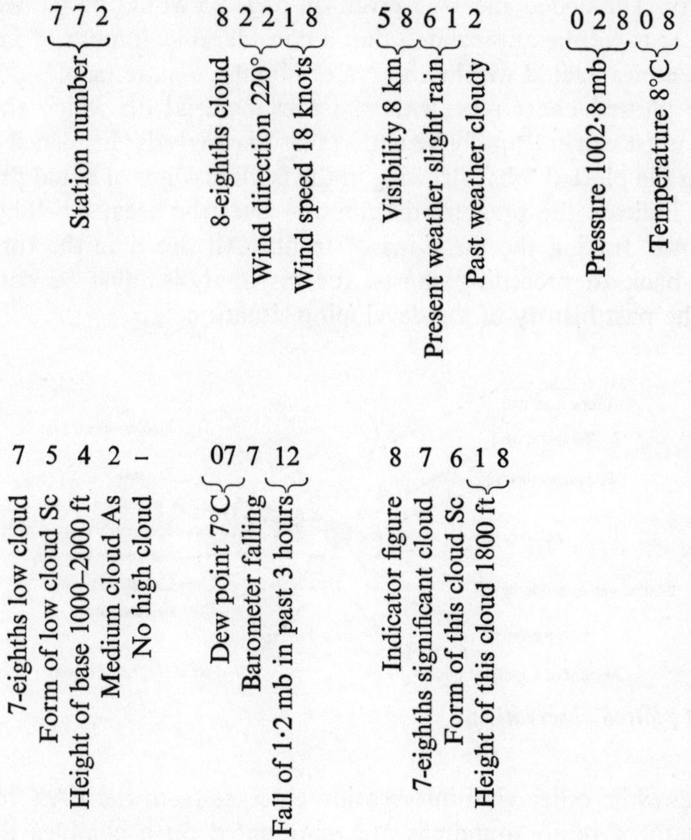

772 — Station number

8 2218 — 8-eighths cloud; Wind direction 220°; Wind speed 18 knots

58 61 2 — Visibility 8 km; Present weather slight rain; Past weather cloudy

028 08 — Pressure 1002·8 mb; Temperature 8°C

7 5 4 2 – — 7-eighths low cloud; Form of low cloud Sc; Height of base 1000–2000 ft; Medium cloud As; No high cloud

07 7 12 — Dew point 7°C; Barometer falling; Fall of 1·2 mb in past 3 hours

8 7 6 18 — Indicator figure; 7-eighths significant cloud; Form of this cloud Sc; Height of this cloud 1800 ft

The long strips of teleprinter paper are hung from clips close to the forecaster's bench: in this way he keeps in touch with current weather events from hour to hour. Not many years ago synoptic charts were plotted from these messages at every office but today this laborious task is normally carried out only at parent stations and of course at the Central Forecasting Office, from which the Western Europe charts are broadcast by facsimile. We see the fully plotted charts slowly appearing on the FAX receiver: a ready-made surface synoptic chart emerges some 70 minutes after the hour to which it refers. A close look shows that the information is in the form of clusters of symbols and figures around small circles which locate the various stations. The plotting technique is based on a standardized 'station model' and employs internationally agreed codes and symbols. At present the charts are still hand-plotted but automated techniques exist and are on the way. The coded message given on page 35 would be plotted as in Fig. 2.1: it will be appreciated that a considerable amount of information is concentrated within an area of about a square inch.

The plotted chart represents the raw material on which the forecaster must work. Rapidly he undertakes his analysis: his pencil weaves among the plotted 'obs', drawing in the isobars (lines of equal pressure) which indicate the pressure distribution over the area, labelling highs and lows, tracing the positions of fronts. All the time the forecaster refers back to preceding charts, for his analysis must be consistent with the past history of the developing situation.

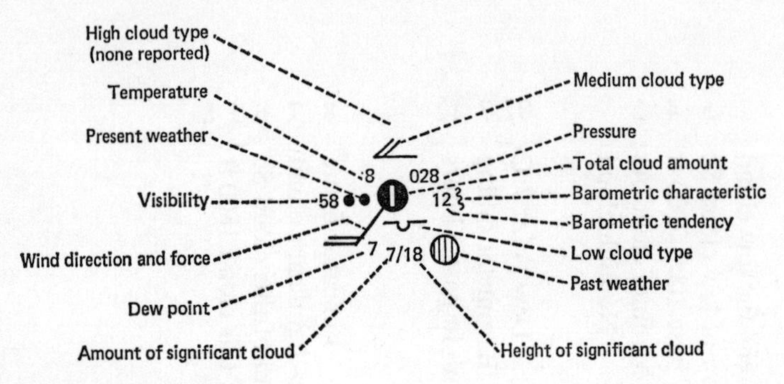

2.1 A plotted observation.

Meanwhile other vital information emerges from the FAX receiver. The data of radio soundings are represented on a complex diagram (the 'tephigram') which depicts the variation of temperature and humidity with height and is a forecasting tool of great value. Upper-air charts show air flow patterns at various heights up to 12 km (40 000 ft), essential information in these days of high-flying aircraft. Especially useful are the forecast charts, the surface pattern ('prebaratic') and various upper-level charts for 24 hours ahead and further surface prognostic charts for 48 and 72 hours ahead which form the basis of extended range predictions. Of especial interest also are the 'nephanalyses', maps showing the cloud patterns as revealed from satellite photographs: these often prove an invaluable check or corrective to other methods of analysis. All this material, together with other teleprinter information, including general guidance reviews and forecasts and special messages like gale warnings from Bracknell and regional forecasts from the parent station, finds its way on to the appropriate hooks all round the office walls.

Supported by all this evidence (and much more when required, such as information en route and destination forecasts for long-distance air flights), the forecaster prepares his story. Forecasts take various forms, depending on the requirements of the airfield. A local area forecast, within 25 to 50 miles radius of the station, is generally issued as a matter of routine at least once daily: this takes care of local short-distance flying. Here the forecaster needs to be well acquainted with the weather peculiarities of his own locality. Longer flights require special route forecasts: these and forecasts for special activities like exercises may necessitate briefings for the crews. The duties of the forecaster do not end here for enquiries by telephone and by personal caller (the 'met. office' is by tradition a hospitable place) must all be answered. Never far from the forecaster's elbow is the Enquiries Book in which all enquiries and replies are duly entered. Sometimes there are special forecasts or warnings to be made known by arrangement to individuals or organizations on or off the station. The 'met. office' can be a very busy place, especially when bad weather is threatening.

CFO

The Central Forecasting Office (CFO) is part of the headquarters complex of the Meteorological Office at Bracknell, Berkshire. Many of its vital functions will be apparent from the preceding section. In the first place, Bracknell is the nerve centre of the national telecommunications network, charged with the day-and-night task of collecting and redistributing with the utmost haste all the data required for forecasting purposes. In a broader context, it is a regional telecommunications hub, receiving reports also from Ireland, Iceland, Greenland, the ocean weather ships, merchant ships and aircraft and making these internationally available: it must also prepare and disseminate pictorial information by land-line and radio-facsimile. The handling of all this information is already highly automated.

The forecasting responsibilities of CFO cover, nationally, not only the preparation and issue of the basic analyses and prognostic charts, supplemented by general advisory material, for the guidance of all other stations in the organization, but also the provision of weather information and forecasts for the benefits of the community at large. Thus the BBC and Press forecasts originate at Bracknell. Notification of special weather conditions such as fine spells, or warnings of hazards like gales, fog, frost or icy roads, are given out for general guidance. Certain organizations – for example, British Rail, the Central Electricity Generating Board, the Area Gas Boards, the local authorities – are warned specifically of impending weather that may affect their operations. Many commercial and industrial undertakings arrange to receive special forecasts for their particular purposes.

Internationally, CFO is a Regional Meteorological Centre, preparing for communication to certain European countries actual and forecast charts covering Europe, part of the North Atlantic and the Arctic. The regional centres are subservient in the hierarchy only to the World Meteorological Centres (of which there are three, Washington, Moscow and Melbourne).

With these responsibilities, CFO plots daily four northern hemisphere charts, eight Western Europe charts and 24 British Isles charts. The upper-air analyses and prognostic charts of various kinds are produced now by computer and the drawing up is automatic. Computer facilities are also employed for research purposes and the vast amount of data constantly received via the communications channels is stored on cards or tape. It is impossible to do justice in a few words to the range of enquiry conducted by the research branches or, for that matter, to the scope of activities under the Directorate of Services.

Weather Information and the Public

It may seem that the energies of our meteorological organization are devoted mainly to satisfying specialized needs but in fact there is a great deal of information ordinarily and regularly available to members of the public with an everyday interest in the weather.

Forecasts of a non-technical nature are issued to all newspapers and press agencies but there is not always much relationship between the information supplied and that actually published. We should not blame Bracknell for the forecasts appearing in some newspapers which are so condensed and generalized as to be practically useless. On the other hand, certain other papers (which deserve our gratitude) print, in addition to tabulated weather reports and detailed forecasts, Bracknell's 'prebaratic' for noon on the day of issue. It should also be remembered that a morning paper publishes a forecast issued the previous evening and this could be out of date by the time it is read.

The forecasts on the four BBC radio channels (and most of the BBC local radio stations) are familiar enough. The Radio 4 bulletins read immediately before the main News broadcasts are more useful than the often over-summarized statements contained within the News itself.

The shipping forecasts broadcast now on Radio 2 (1500 m) are of a rather more technical nature and contain some synoptic information, which make them useful for exercises in amateur analysis and forecasting (see Chapter 11). The television forecasts on BBC present both actual and forecast charts with a symbolic representation of the expected weather, though only the late-night broadcasts on BBC 1 give time enough to allow a real appreciation of the weather situation. The TV weather-men from the London Weather Centre have done much to create a weather-conscious public.

For the most up-to-date information and forecasts, Weather Centres after the style of the London Weather Centre at High Holborn (formerly in Kingsway) are now open in Manchester, Newcastle, Southampton and Glasgow for personal enquiries on weekdays, and some 30 other stations are currently designated to act as public service offices. Their telephone numbers may be found in area telephone directories and there is no cost to the enquirer beyond that of the call. Beginning in 1956 in the London area, the automatic telephone weather service (ATWS) has expanded steadily and now provides tape-recorded forecasts (changed at least every six hours) for many local areas: again the telephone directories give the necessary details. There is also a wide range of special services available at a cost to the enquirer whose needs are not met through the usual channels. Information may be sought from the Director-General, Meteorological Office, Bracknell, Berks.

The amateur who likes to follow the march of weather events or the teacher in school or university who wishes to use synoptic data for teaching or research purposes will find invaluable sources in the *Daily Weather Report* and the *Daily Aerological Record.* Both publications are printed at Bracknell and can be supplied by post to any address on application to the Meteorological Office: there are special rates for educational establishments.

The *Daily Weather Report*, in its present form, consists of four pages. Two are covered with detailed weather reports, for 0000 hr and 0600 hr GMT of the day of issue and 1200 hr and 1800 hr of the day before, from 55 British Isles stations and a number of ocean weather ships. The other two pages carry, for the same times, four synoptic charts, of which one (noon) is a small-scale chart covering most of the northern hemisphere and the others, of larger scale, cover Western Europe. Figure 7.2 is based on the northern hemisphere chart of a *Daily Weather Report.* Included also is a brief statement of the general synoptic development and an explanation of the symbols and figures used on the charts. The *DWR* is posted in the afternoon and arrives at most destinations next morning.

Although it necessarily reaches us a day late, the *DWR* offers the explanation of past weather and the background of present developments against which we can check our own observations and whatever forecasts we may see or hear. We can also use the station data to prepare past charts in as much detail as we wish. To plot charts giving essentially the same information as those used by the Meteorological Office is feasible but requires knowledge of the codes and symbols: this may be obtained from the quarterly *Introduction to the DWR* or from the booklet *Instructions for the Plotting of Weather Maps* (Met. O. 515, from HMSO). The *DWR* charts themselves are plotted using a simplified station model, which gives an appropriate amount of information for this scale of map. This is sufficient for elementary teaching purposes and the examples of synoptic charts used later in this book have been prepared in this style. The plotting model and an explanation of the symbols and figures likely to be encountered are given on page 42. A map showing the station data alone but including the pressures is clearly suitable raw material for exercises in simple analysis (drawing in of isobars, insertion of fronts, etc). The station pressures can be obtained crudely by interpolation from the isobars on the *DWR* map or precisely from the station reports: the column headed PPPTT gives the last three figures of the pressure in millibars (as well as the temperature), e.g. 98706 means pressure 998·7 mb (and temperature 6°C) and 21318 means pressure 1021·3 mb (and temperature 18°C).

The companion *Daily Aerological Record* may seem of more remote interest, especially as it is issued a week or more in arrears. It is a six-page report including upper-air observations from the British radio-sounding stations and the nearest ocean weather ships. This infoımation enables us to plot 'tephigrams' if we wish to master the intricacies of this diagram (copies are obtainable from the Meteorological Office) or much simpler temperature-pressure graphs or the temperature-height diagrams which are used in this book. It is again necessary to become familiar with the codes which are explained in the *Introduction to the Daily Aerological Record.* For very approximate temperature-height graphs, Fig. 2.2., based on an 'average atmosphere', will allow conversion from any pressure in millibars to a round number in either metres or feet. The upper-air charts may seem far removed from our view of weather from the ground. Sufficient to say at present that they are of vital importance in modern meteorology, that they consist of

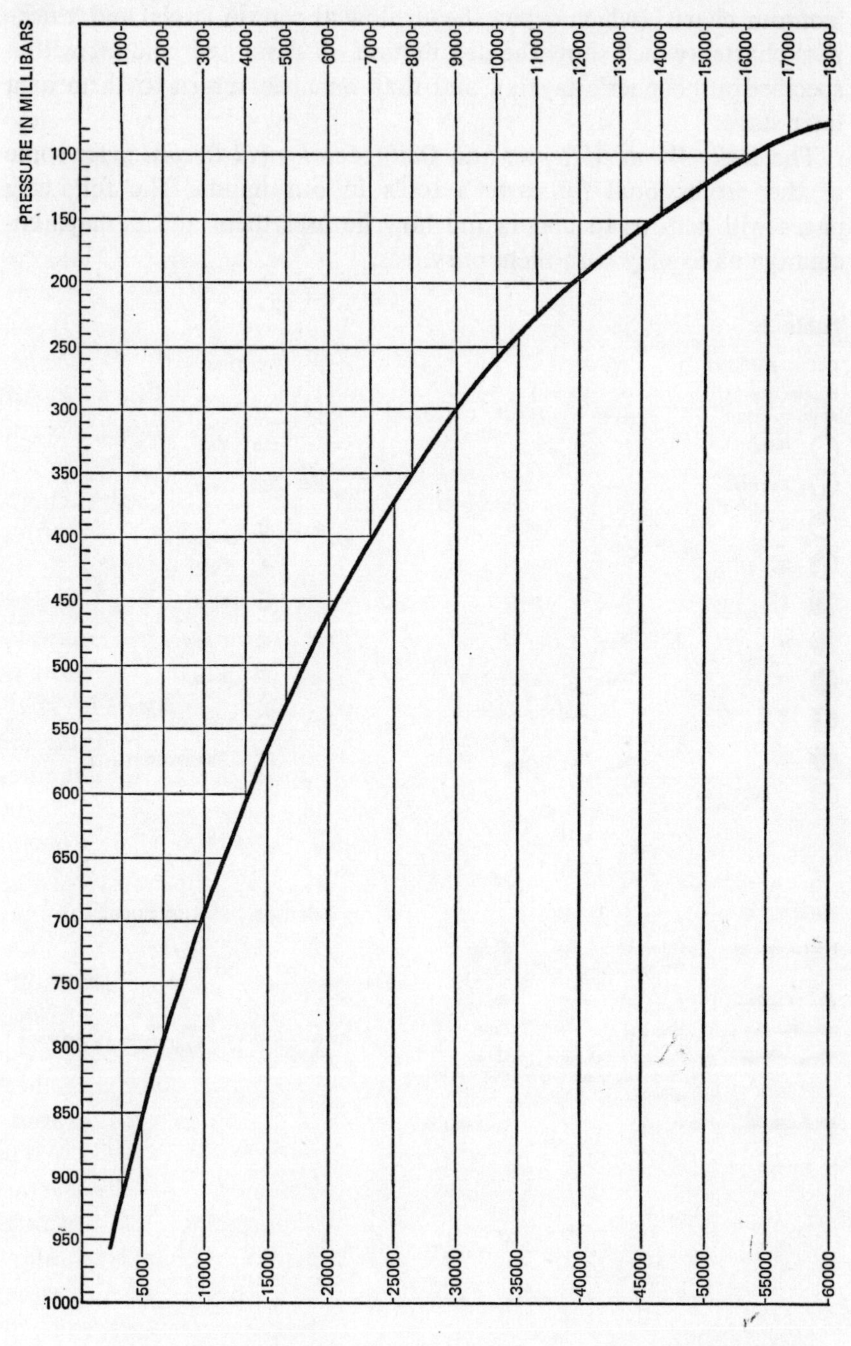

2.2 Relation between pressure and height in an 'average atmosphere'.

'contour charts' (which depict the air flow at certain levels) and 'thickness charts' (which show the distribution of warm and cold air within specified atmospheric layers), and that we must return to them at a later stage.

The *Daily Weather Report* and *Daily Aerological Record* place some of the professional forecaster's tools in our hands. The following pages will help us to understand how he uses them and perhaps encourage us to play with them ourselves.

Table 3

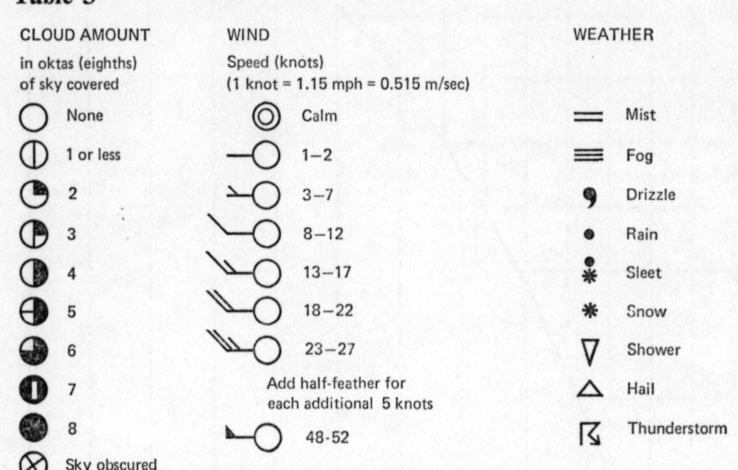

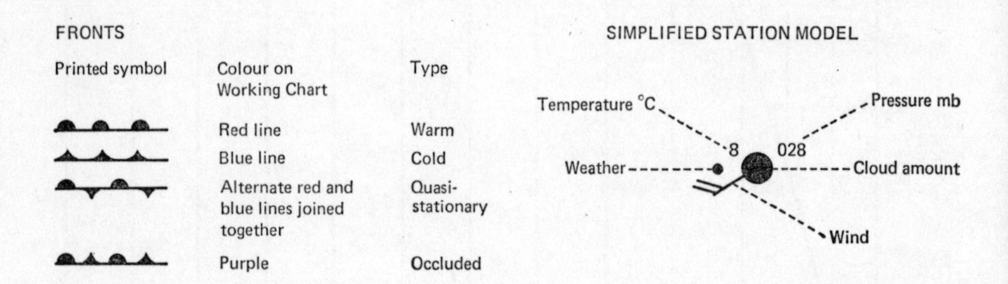

3 The Causes of Weather

There is no obvious answer to the question 'What causes weather?' even though this is the most obvious of questions, which sooner or later any intelligent child will ask. The atmosphere is a complex and interlocking system, bristling with feedbacks, so that it is difficult to disentangle cause and effect. We have to accept that, with a planet of the size and history of ours, rotating as ours does daily about its axis and yearly about the sun, maintaining as it does an atmosphere of a certain composition and resultant properties and having a particular surface configuration of land and sea, there is associated, in ways that are by no means completely understood, a certain pattern of weather and climate. What we can recognize as a starting point is the need for a vast input of energy to power the atmospheric machine.

This energy comes from the sun, in *radiant* form, much of it perceptible to us as light and heat. Even the light and heat we obtain from burning coal or oil come from sun-derived supplies, stored in the rocks over many millions of years. Solar energy is basic to our considerations. Energy cannot be destroyed but its various forms are interconvertible under certain conditions. Such transformations readily occur in the atmosphere, so that incoming solar energy may be converted into the energy of motion of moving air masses or the electrical energy of thunderstorms. We shall see that practically all the atmospheric processes and conditions that make up the weather depend ultimately on solar energy and the way it is deployed over different parts of the earth's surface. In a real sense, this is the first cause of weather.

Energy in Transit

We need at this stage to tie some technical labels to methods of energy transference which are familiar enough in everyday life, but which have special relevance in meteorology. *Conduction* is a transfer of heat energy by contact from hotter to colder particles, as when a poker left stuck in the fire becomes hot at the handle end. The heat of a body is a manifestation of the energy of vibration of its constituent molecules: in conduction along the poker this energy is communicated from the more agitated molecules of the hot end to the more sluggish ones of the handle. Conduction is thus most important in solids, in which the molecules are tightly held together, but less effective in liquids and least so in gases. As a process it is negligible in the atmosphere generally but becomes important at and near the common surface of ground and air.

In liquids and gases, mass flow is possible and heat may be transported from hotter to colder regions by the actual movement of the fluid. This is *convection*, clearly of great importance in the mixture of gases that constitutes the atmosphere. Meteorologists usually reserve the term for vertical motions and distinguish horizontal movements as *advection*. Associated with both is heat transfer by movement of water in the atmosphere: this we shall consider later in the chapter.

No heat transfer mechanisms so far mentioned can be responsible for the transmission of energy from sun to earth, since they require matter in which to operate and practically all the 93 million miles separating these two bodies is effectively empty space. Only *radiation* can transfer heat through matterless space and the sun is a most effective *radiator*. The everyday use of this word is confusing. The vanishing open coal fire warmed us (or half of us) by radiation, clearly not by convection, since most of the warmed air escaped regrettably up the chimney. A so-called 'radiator', however, performs its task rather little by radiation and much more by convection, by warming the adjacent air and setting up a circulation throughout the room.

Nevertheless, the 'radiator' does radiate as, in fact, all bodies do. Physicists regard radiation as an energy emission in the form of waves (analogous to sound waves in air or ripples on water), which are conveniently described in terms of their speed of travel, the number of waves per second (*frequency*) and the distance between successive wave crests (*wave-length*). Light, heat, radio waves, X-rays, are all waves of this kind, having the same velocity (the 'speed of light', 186 000 miles/sec or 3×10^{10} cm/sec) but differing in frequency and wave-length and in their effect on objects placed in their path. The idea of wave-length is familiar to us from the dial of our radio sets:

wireless waves (which are very long) do not affect us but can be picked up by tuning a suitably constructed electrical circuit. If we imagine our radio dial extended enormously towards the shorter wave-lengths, we would come eventually to a range within which radiation affected us physically by warming us (radiant heat) and, shorter still, by warming us and also affecting our eyes as light (visible radiation). At the extreme short-wave end come emanations such as X-rays and cosmic rays. This is the so-called *electro-magnetic spectrum* (Fig. 3.1).

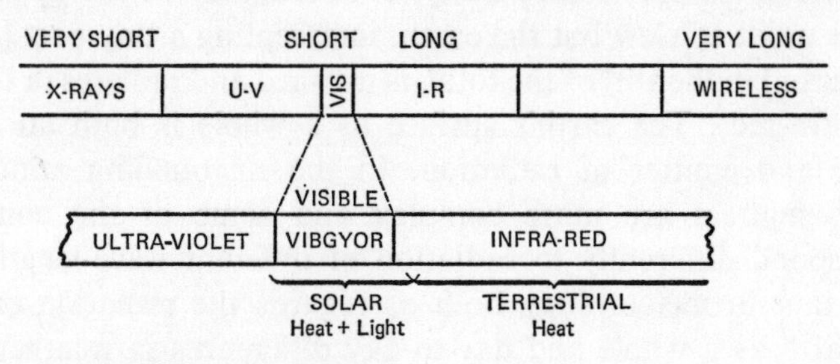

3.1 The electro-magnetic spectrum.

Meteorologists are concerned with only part of the spectrum, the relatively long heat waves and the shorter ones effective as heat plus light. One of the laws governing radiation lays down that the hotter the body is, the shorter the wave-length of its maximum radiation. Thus the glowing coal fire or electric bar fire emits short-wave radiation whereas the much cooler 'radiator' sends out only invisible, long-wave heat radiation. Similarly, the sun, with a surface temperature of nearly 6000°C (about 11 000°F) emits short-wave radiation, with a maximum intensity within the visible spectrum (light) but also including invisible waves shorter than the violet (the 'ultra-violet') and longer than the red (the 'infra-red'). In contrast, the earth's surface and atmosphere radiate at wave-lengths appropriate to their mean temperature of around 15°C (say 60°F), which is to say entirely within the infra-red (Fig. 3.1). Another, more obvious, radiation law states that the hotter a body is, the more intensely it radiates, so that solar radiation is expectedly much more intense than earth (terrestrial) radiation.

What happens to an object placed in the path of a radiation beam depends on the physical properties of the object. A body may *transmit* radiation, just as window glass allows the passage of light, when neither the radiation nor the transmitting body are affected in any way. Other bodies *reflect* radiation, as a mirror reflects light: in this case, the reflecting body remains unaffected and the radiation is altered only in the direction of its travel. Still other bodies *absorb* radiation, which means that the radiant energy becomes converted into the internal (heat) energy of the absorbing body. This is an energy transformation and the temperature of the absorbing body increases. It is in this way that the earth's surface is warmed by solar energy, but it must itself radiate in wave-lengths determined by the temperature it achieves.

To sum up, the sun emits short-wave radiation, much of it in the visible part of the spectrum. Nearly all of this radiant energy is wasted from our point of view but the earth, intercepting a tiny part (about a two thousand millionth) of the total, is warmed and radiates in the long-wave (infra-red). The earth's surface as a whole is both an efficient absorber and emitter of radiation. In the surrounding atmosphere, however, matters are more complex and some of the constituent gases respond differently to radiation of different wave-lengths. This is important in meteorology both as regards the radiation exchange of the earth as a whole and day-to-day differences in weather over a particular spot.

Radiation and the Atmosphere

Measurements from high mountain observatories and more recently from satellites have established an average value for the rate at which energy is received at a point just outside the earth's atmosphere. However, this value (the so-called *solar constant*) is of less interest to us than the solar radiation received at the earth's surface after having passed through the atmosphere: this we call *insolation*. At any point on the earth's surface, insolation depends on the solar constant; on astronomical factors such as the distance between earth and sun (remembering the elliptical orbit), the elevation of the sun (the higher the sun, the greater the insolation) which varies with latitude, season and time of day; and on the rather variable condition of the atmosphere.

The atmosphere is a mixture of gases, of which some (like oxygen and nitrogen) are everywhere present in fixed proportion, while others vary in concentration with time and place: these include water vapour and carbon dioxide, which are mainly concentrated in the lower layers, and ozone, which is most important at great heights. Solid impurities like dust, smoke and salt, are found mainly in the lower layers. Most of these constituents reduce incoming solar radiation in various ways.

About a seventh part of the sun's energy is absorbed in the atmosphere, mainly by oxygen and ozone at high levels and a little by carbon dioxide and water vapour nearer the earth's surface. A larger part is reflected by water droplets, particularly from the tops of clouds, and largely lost into space. Thick clouds reflect nearly four-fifths of the radiation falling upon them, thin broken clouds much less. Snow-covered portions of the earth's surface also contribute to the loss by reflection, but the major factor is the amount of cloud. The earth as a whole is on average something like half-covered by cloud, which means that, over all, about two-fifths of the incoming solar radiation is lost by reflection. This average value (about 40%) is referred to as the *reflectivity* or *albedo* of the earth.

Dust and smoke particles, which are smaller than water droplets, *scatter* rather than reflect, sending the radiation off (unaltered) in all directions. Scattering is more effective for the shorter wave-lengths so that the bluer light is more easily scattered than the longer red rays. Some of the scattered light is lost into space but some is directed downwards as a diffuse radiation from the sky in general. A cloudless sky appears blue because we see it by diffuse light which has undergone much scattering and is rich in the blue rays. On the other hand, when we see the sun through a smoke haze, or through a great thickness of atmosphere when it is low in the sky, it looks reddish, since the light has been robbed of much of its blue and is correspondingly richer in the red rays.

Subject to these erosions, less than half the incoming solar radiation on average survives its passage through the atmosphere and is finally received and absorbed by the earth's surface. Equally noteworthy is the comparatively small fraction of the available solar energy that is actually absorbed by the atmosphere (and that mainly at great heights). We express this in the statement that the atmosphere as a whole is largely *transparent* to solar radiation. It follows from this that the atmosphere is warmed very little by direct solar energy.

But the atmosphere responds quite differently to the radiation emitted by the earth's surface, absorbing strongly in these longer wave-lengths. This means that while it is true that the sun is the ultimate source of its energy, the atmosphere is heated directly from the earth's surface, not from above but from below. This is of great importance not only for the overall atmospheric heat budget but in connection with individual air masses, which acquire their temperature characteristics from underlying surfaces.

Of the constituents of the atmosphere, water vapour, carbon dioxide and ozone are important absorbers of long-wave radiation, absorbing in most of the terrestrial wave-lengths (but not in all, so that there is always some direct loss of radiation into space through these 'windows' in the spectrum). Water is especially significant. Absorption continues even in cloudless skies, for there is always some water vapour present: in moisture-laden cloudy conditions, absorption is even more effective. Cloud surfaces, with such a high albedo for solar radiation, reflect hardly at all in these longer wave-lengths. The atmosphere, warmed by this absorption, acquires a temperature of the same order as that of the earth's surface (though lower) and must itself radiate in the same long wave-lengths. Part of this radiation is emitted upwards and lost into space but part is directed downwards to the earth's surface. Here it is again absorbed, converted into heat and re-radiated. Radiation and counter-radiation thus proceed continuously between earth and atmosphere.

The Heat Budget

The earth's surface and atmosphere are, in the long view, growing neither hotter nor colder. This means that all the energy deriving from the sun and being converted into heat must ultimately be lost by terrestrial radiation: there must be an over-all balance between energy income and outgo. In the following budget account of these complex energy transactions, average conditions are assumed for the earth and atmosphere as a whole and for the year as a whole. It has to be an approximate and tentative statement and not too much significance should be attached to the precise values assigned to the various items, since the estimates and computations are constantly under review.

The processes involved are depicted diagrammatically in Fig. 3.2. The left-hand side of the diagram is concerned with short-wave radiation, showing that of 100 units of solar energy arriving at the outer limit of the atmosphere, depletion by the various means already discussed allows only about 45 to reach the earth's surface. Of this about one-third arrives as diffuse sky radiation.

The right-hand side of the diagram shows how the energy absorbed by earth and atmosphere is eliminated, so preserving the heat balance. Of the radiation emitted by the earth's surface, the major part is absorbed and returned by the atmosphere: however the effective radiation loss – the excess of outgoing over that received – eliminates over half the 45 units absorbed from the sun. Of this terrestrial (long-wave) radiation, about a third is lost directly into space (through the atmospheric 'windows') and the rest re-absorbed by the atmosphere. About as much heat is lost to the earth's surface by evaporation, carried upwards in latent form and later yielded to the atmosphere by condensation: the mechanisms of these processes we shall examine shortly. Conduction and convection in themselves play only a small role in the effective heat exchange, since the heat flux may be upwards or

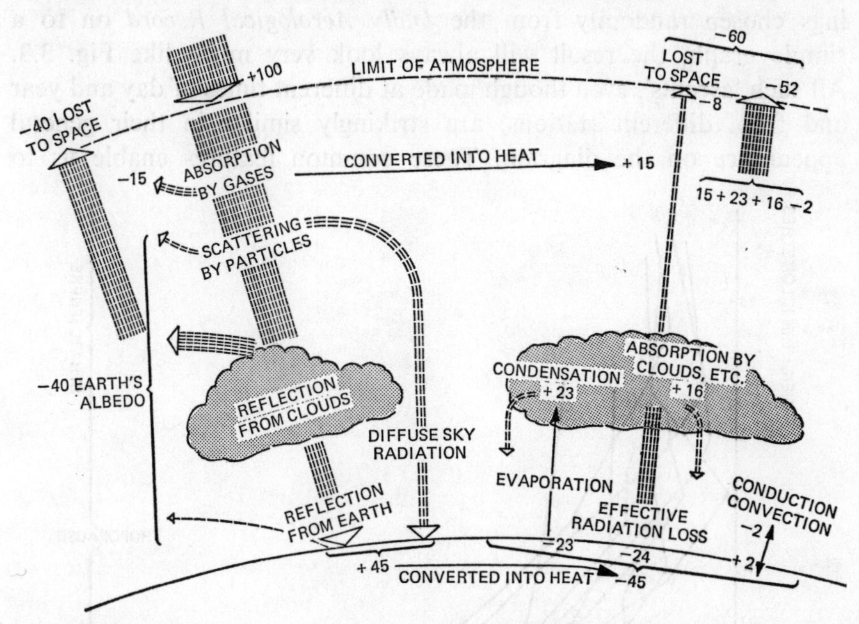

3.2 Diagrammatic representation of the heat balance of the earth and atmosphere.

downwards according to circumstances: some authorities agree on a small net downward transfer. Finally the diagram shows that the heat directly absorbed from solar radiation, plus that absorbed by clouds, etc, plus that resulting from condensation, less that allowed for conduction and convection, remains to be radiated by the atmosphere back into space. In something like this manner, we assume the books are balanced.

It remains to draw attention once again to the selective response of the atmosphere to the radiation it receives. Because the atmosphere absorbs much of the outgoing terrestrial radiation and returns part of this to the earth, it becomes clear that the earth enjoys an extra bonus of heat due to this protective function – sometimes known as the Greenhouse Effect of the atmosphere. It has been calculated that, without its atmosphere, the mean temperature of the earth's surface would be some 25°C (45°F) colder than it actually is and the daily and seasonal extremes of temperature would be considerably greater.

Temperature in the Vertical

If we plot the variation of temperature with height from radio-soundings chosen randomly from the *Daily Aerological Record* on to a simple graph, the result will always look very much like Fig. 3.3. All such 'ascents', even though made at different times of day and year and from different stations, are strikingly similar in their general appearance on the diagram. These common features enable us to

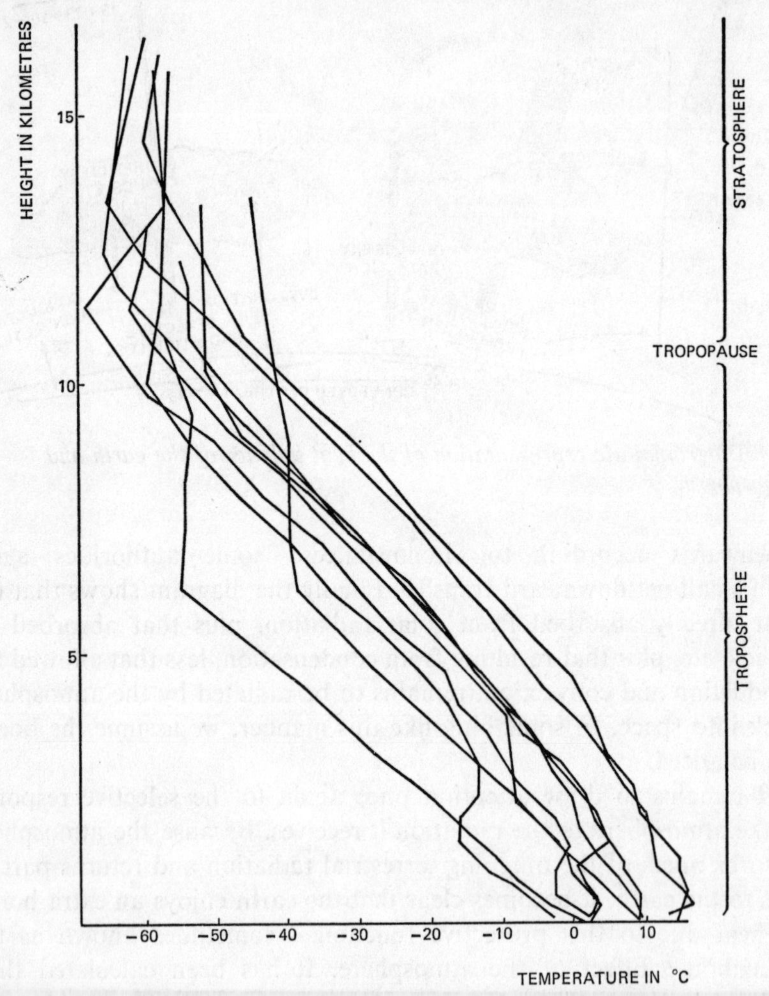

3.3 Some typical radio-soundings.

identify at least three atmospheric layers, distinguished by the slope of the curve, which represents the behaviour of temperature with height. The greater part of each curve shows a fall (lapse) of temperature with height and the rate at which this decrease occurs is called the *lapse rate*: this is usually expressed as a fall of so many degrees C per 100 m (or degrees F per 1000 ft). A negative lapse rate implies an increase of temperature with height, which is more often known as a *temperature inversion*. A layer within which temperature remains the same is an *isothermal* layer.

In these terms we can distinguish in the curves of Fig. 3.3 a lower atmospheric layer in which temperature decreases generally with elevation – the *troposphere* – and an upper layer which is not far from isothermal or tends towards a negative lapse rate – the *stratosphere*. The level at which the lapse rate changes has been christened the *tropopause*: sometimes the transition is sharp, at others it occurs over a zone of some thickness. The average height of the tropopause in middle latitudes is around 10–11 km (or, say 30 000–35 000 ft) but, as the diagram shows, the actual heights can vary widely. Apart from this the various curves differ most in the lower troposphere, where temperature inversions are evident in some cases, but look remarkably similar as regards their slope in the middle and upper troposphere sections. The average lapse rate in the troposphere as a whole, calculated from many temperature soundings, is about 0·65°C/100 m (approximately $3\frac{1}{2}$°F/1000 ft).

The thermal structure of that part of the atmosphere depicted in Fig. 3.3 is very roughly what we should expect from the nature of the heat balance, since the earth's surface is, in general, the warmest region and the atmosphere, gaining heat from below but losing it into space, must be warmer with increasing proximity to the surface. The exact slope of the typical temperature-height curve also reflects the importance of convection currents which effectively stir a certain thickness of the atmosphere and produce, by means we shall later consider, the uniform lapse rate expressed by the parallelism of part of the curves. The tropopause is visualized as the maximum height to which these convection currents penetrate and therefore, since their energy source is the warmth of the surface, the effective upper limit of the influence of the earth on the temperature of the atmosphere. Over tropical regions, where surface heating is most intense and convection currents can extend to greatest heights, the tropopause is at its highest, at about 17 km (in round figures 55 000 ft) and the troposphere therefore thickest: whereas over the poles, where the least energy is available for convectional activity, the tropopause is lowest (around 8 km or 25 000 ft).

The increase in temperature or absence of further lapse in the lower stratosphere shown in Fig. 3.3 is due to strong absorption of solar radiation by ozone, which exists in relatively high concentration at these levels. The ozone concentration itself varies with latitude, being greater above the polar regions than above the equator, so that the stratosphere is warmer above high latitudes than above low, in contrast to conditions in the lower troposphere. The stratosphere is not, as was once thought, a region of still and uniform conditions. Rather slow vertical movements take place there and often strong horizontal winds. However, the stratosphere is extremely dry and, since weather as we know it is much associated with moisture, there is support for the popular notion that the stratosphere is 'above the weather'. Beyond the stratosphere are found further layers with different properties but these again do not appear to have any demonstrable connection with the weather we experience at the base of the troposphere.

Temperature over the Surface of the Earth

The distribution of temperature horizontally over the surface of the earth is largely, as we should expect, a response to the variation of insolation with latitude. Low latitudes, where the mid-day sun is always high and insolation strong, are in general the warmest and temperatures decrease towards the polar regions where the rays from the low sun fall obliquely and have little intensity. Figure 3.4 shows the computed variations with latitude of three significant values, the average solar energy receipt at the top limit of the atmosphere (curve *A*), that actually absorbed by the earth-atmosphere system (curve *B*) and the average outgoing (long-wave) radiation (curve *C*). The difference between curves *A* and *B* represents the earth's albedo and it can be seen how this in fact varies with latitude, being rather less than 40% near the Equator but much more at the Poles. Curve *C* expresses the fact that low latitudes, being warmer, radiate more than do higher latitudes. Of most interest, however, is the relation between curves *B* and *C*, which show that incoming radiation exceeds outgoing in low latitudes, while the reverse is true of middle and high latitudes. The two curves cross at about latitude 35°, which in terms of area approximately divides a hemisphere into halves. While an over-all energy balance is maintained for the earth-atmosphere as a whole, there is a definite unbalance in the heat exchange at different latitudes.

If low latitudes are heat *sources* and middle and high latitudes heat *sinks*, we may ask why the former are not becoming steadily warmer and the latter steadily colder. The answer is inescapably that some mechanism exists for transferring heat from the regions of excess to those of deficit and we shall seek it (in Chapter 7) in the complex motions that constitute the general circulation of the atmosphere and, secondarily, in the movement of ocean water.

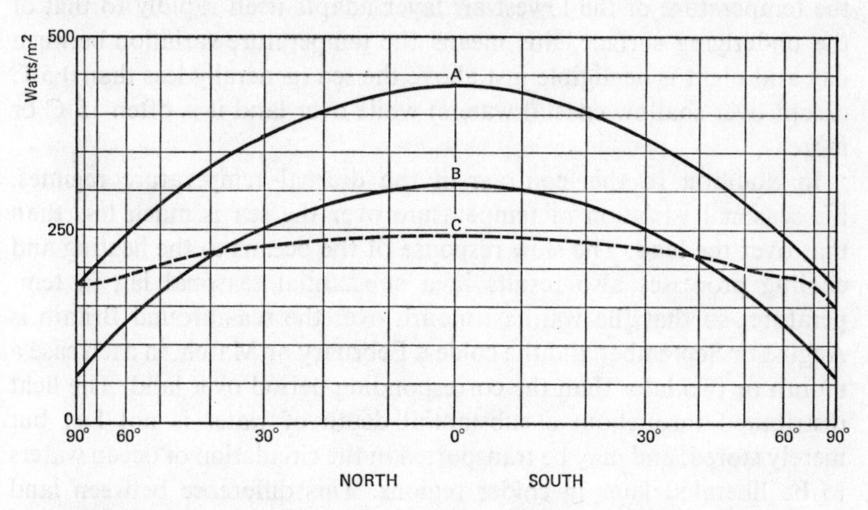

3.4 The unequal heating of the earth and atmosphere. Curve A, solar radiation received at the top of the atmosphere. Curve B, solar radiation absorbed by earth and atmosphere. Curve C, terrestrial radiation leaving earth and atmosphere. (After Sellers.)

The temperature conditions at a given location, however, depend on more than the radiation exchange. A very important factor is the nature of the surface on which the insolation initially falls, and the most significant contrast in this respect is that between land and water surfaces. The average *specific heat* of the solid ground is about one-third that of water, which means in ordinary language that it needs three times as much heat to raise the temperature of a quantity of water through so many degrees as is required to heat the same quantity of solid earth through the same number of degrees. The solar radiation falling on the land is immediately absorbed in a very thin surface layer, from which conduction downwards proceeds slowly and only to shallow depths: in short, the heating effect is highly concentrated and the rise of temperature is therefore considerable. On the other hand, radiation penetrates relatively deeply into water and, in further contrast to the solid ground, the mixing and stirring which are possible in a fluid spread the heating effect over a considerable depth. It follows that the rise of temperature at any point is very small.

The result of these contrasted properties is that, faced with the same amount of solar heating, water responds only slightly but land considerably. Similarly, at night when outgoing radiation dominates, land temperatures drop sharply, water temperatures very little. Since the temperature of the lowest air layer adapts itself rapidly to that of the underlying surface, this means the temperature variation between day and night is negligible just above the sea (generally less than 0·5°C except over shallow coastal waters) while over land it is often 15°C or more.

In addition to this contrast in the diurnal temperature regimes, the seasonal variation of temperature over the sea is much less than that over the land. The slow response of the oceans to the heating and cooling processes also results in a substantial seasonal lag in temperature, so that the warmest month over the seas around Britain is August or September and the coldest February or March, in each case a month or two later than the corresponding period over land. The heat distributed throughout a substantial depth of water is not lost but merely stored, and may be transported in the circulation of ocean waters to be liberated later in colder regions. This difference between land and water profoundly modifies the general atmospheric circulation and introduces important differences between the so-called maritime and continental climates.

To-day's Temperature

At this stage we can forget average conditions for the moment and consider day-to-day temperature fluctuations, at a given spot, which are largely dependent on the state of the atmosphere. With clear daytime skies, the depletion of incoming solar energy is at its least and probably two-thirds of the possible radiation is absorbed at the surface. Over land areas in the absence of strong winds (which, as we shall see, are unfavourable to temperature extremes) the highest maximum temperatures possible for the latitude, the time of year and the type of air are realized. A cloudless night will allow the greatest loss of heat by terrestrial radiation and low minimum temperatures must therefore be expected (again assuming quiet conditions). With clear skies, the air temperature rises from a minimum about dawn to reach a maximum around mid-afternoon, that is, two or three hours beyond the time of maximum insolation (noon). This lag between temperature and sun is explained by the fact that up to 2 or 3 p.m. the gain by insolation still exceeds the loss, but thereafter the balance tips the other way, outgoing radiation overtakes the diminishing income and the temperature falls.

With cloudy skies, incoming radiation is cut down to a minimum: most of it is reflected from the cloud tops and only about one-fifth is absorbed at the earth's surface. Under these conditions temperatures rise only slightly and low maxima are recorded. Cloudiness is however an advantage during the night when outgoing terrestrial radiation is reduced through absorption by, and back-radiation from, the cloud cover. The effectiveness of this blanket depends largely on the height of the cloud: a complete cover of low cloud reduces the nocturnal cooling to about an eighth of what it would be with a cloudless sky, but very high cloud has only a small effect. With this exception, cloudiness at night tends to maintain relatively high minimum temperatures. It follows from these considerations that clear skies both day and night favour temperature extremes, that is, a large diurnal variation. These are the conditions in which, especially in air of cool origin, the pleasant sunny afternoon bears a warning of the chill evening to follow and which cause anxiety to farmers and horticulturalists in the marginal spring and autumn seasons when a low minimum may mean a damaging frost. Continuous cloudiness makes for a small diurnal range. Other combinations are, of course, possible; we can readily see that clear days followed by cloudy nights give the best possible temperature regime, while with cloudy days and clear nights we experience the terrestrial heat exchange at its least advantageous.

Cloudiness not only reduces the diurnal range but may also blur the times of the temperature extremes. For example, a spread of heavy cloud during the late morning may advance the time of maximum temperature to mid-day or earlier. The replacement of one air mass by another, with the accompanying temperature change, may completely upset the normal diurnal march of temperature, even to the extent of producing a maximum at night. The daily variation depends also on the wind strength. Only still or gently moving air is subject to great temperature extremes, cloud permitting: stronger winds, because of the accompanying turbulence (see Chapter 4), diffuse the effects of heating and cooling upwards through a fair thickness of atmosphere, which makes for a reduced range at the surface. Thermograph traces illustrating certain of these conditions are reproduced in Fig. 3.5.

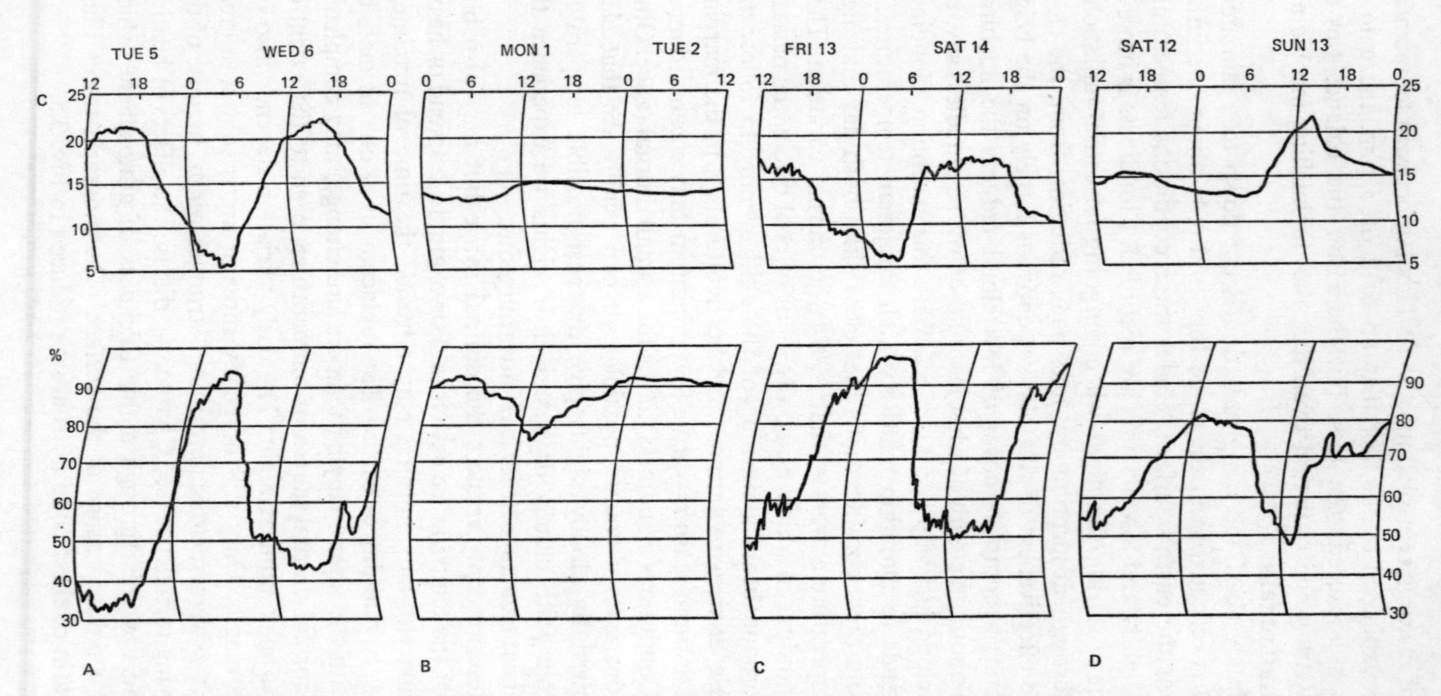

3.5 Some typical thermograph and hygrograph traces (a) Clear skies night and day (b) cloudy night and day (c) clear night followed by cloudy day (d) cloudy night followed by clear day.

Water in the Atmosphere

The existence of water in the atmosphere is a major element in any review of the basic causes of weather and, as already indicated, it plays an important rôle in the atmospheric heat balance. Water is present in the atmosphere in all three states of matter, gaseous, liquid and solid, in other words, as vapour, droplets and ice particles. These are linked in what is sometimes known as the *hydrological cycle*. For convenience we shall consider as the initial stage the process by which moisture enters the atmosphere from underlying surfaces: this of course is *evaporation,* which takes place readily not only from water bodies but also from many soil and vegetation-clad surfaces. Evaporation is a phase change from liquid to vapour but may be taken in this general context to include *sublimation*, the direct change from solid to vapour, which may occur from snow and ice surfaces.

In the atmosphere, water vapour returns to the liquid phase in the process of *condensation*, forming the droplets which make up clouds. Many clouds also contain ice particles, so that freezing must be recognized as yet another phase change that can take place in the atmosphere. Clouds are a fascinating part of the weather we observe and the different forms they assume often reveal much of the mechanism of their origins. There is no need to regard all clouds with suspicion: some are pleasing accompaniments of fair or fine weather. Others are associated with the third phase of the circulation of water in the atmosphere, its return to the earth's surface as *precipitation.* In meteorology this term is used to denote all water deposits derived from the atmosphere (including, for example, dew and hoarfrost): for the moment, however, we shall concern ourselves only with what is literally 'fallout' from clouds as rain and snow, two forms which, as we shall see, are often related. Movements of water within the ground and via the natural drainage systems into the seas and oceans comprise the non-meteorological elements that complete the hydrological cycle.

Evaporation

It so happens that everyday experience teaches us the various factors that affect the rate of evaporation. The higher the temperature of the water body, the more evaporation will occur: we apply heat in order to boil a kettle of water and evaporation occurs, though as an unintended by-product of the exercise. There is much more evaporation from warm seas than from cold. Evaporation rates also depend on the moisture content of the air above the water surface, for air has only a limited capacity for holding water in the vapour state: in still, moist, muggy air, wet streets or damp washing take a long time to dry. A third important factor is the wind strength. With fresh winds and the accompanying eddying and swirling (turbulence), there is constant renewal of the air close to the evaporating surface, so that successive volumes of drier air arrive to take up their quota of water vapour before moving on. In this way, a good deal of moisture can be evaporated, as every housewife knows.

Now part of the atmosphere, the water vapour is spread upwards by convection or horizontally by advection, to influence weather events perhaps far removed from the initial evaporating surface. But the process of evaporation itself affects the heat exchange between earth and atmosphere. As a mechanism, evaporation needs heat: energy is required to free the water molecules from the tighter bonds of the liquid state. We see this illustrated daily again when we boil water: its temperature rises to boiling-point and remains there even though heat is still being applied. This further heat, which is used solely to convert liquid water to water vapour, is the *latent heat* of vaporization of water. In the atmosphere, the necessary latent heat is taken from the internal heat content of the evaporating water body so that both this and the air in immediate contact with it suffer cooling by evaporation.

Condensation

The heat thus acquired is now part of the internal energy store of the water vapour and is transported with it. But when condensation occurs, the water then returning to the liquid state, the latent heat is released and serves to warm the air. Thus the atmosphere eventually gains the heat originally lost from the earth's surface, with important implications for the over-all heat budget as we have seen.

Condensation occurs only when certain conditions are fulfilled. The first of these is that the air must be *saturated*, that is incapable of holding any more water in the vapour state: as we noted earlier, its capacity for doing so is strictly limited. There are two ways of inducing saturation in air. One is to go on evaporating water into a sample of air until it is saturated, when some of the water vapour condenses out: a condition of balance is reached when as many water molecules leave the air and enter the liquid as move in the opposite direction. This method of saturating air is of only limited importance in the atmosphere, though it may well be effective in a cold bathroom when a hot bath is running. The second way rests on the fact that the capacity of air for containing water vapour depends on its temperature and is greater the higher the temperature. On the other hand, cooling a sample of air reduces this capacity and brings it eventually to saturation point. This is the most important means by which saturation and subsequent condensation are achieved in the atmosphere: we must later examine in some detail the various processes that result in the cooling of air.

Laboratory experiments have shown that it is extremely difficult to induce condensation in air that has been cleaned of all impurities: under such (highly artificial) conditions, water vapour can exist in a *super-saturated* state. A second essential requirement for the condensation process seems to be the presence in the air of minute particles or droplets – known as *condensation nuclei* – around which water droplets can form. The most likely substances to be effective as condensation nuclei are those that are *hygroscopic*, i.e. have an affinity for water. It happens that the atmosphere is always sufficiently supplied with such nuclei, although their nature and concentration vary from place to place. They are very small, mainly of the order of a micron (0·0001 cm) in diameter, and are found in concentrations of some thousands to many hundreds of thousands per cubic centimetre. Ordinary sea salt seems the most common: salt particles enter the atmosphere in sea spray, being left in suspension when the water evaporates. Certain products of the combustion of fuels also act as nuclei, including sulphuric acid droplets (forming from sulphur dioxide and water) and these nuclei are most concentrated over industrialized urban areas where their abundance is thought to encourage the premature formation of fog or of thicker fog than over adjacent, cleaner rural areas.

Precipitation

The difference between the droplets that float around as cloud and the drops that fall out as rain is one of size and weight only. Cloud droplets, with a mean diameter size of 5 to 10 microns, are very light and remain suspended in the air or would sink so slowly that evaporation would destroy them long before they could reach the ground: only the very largest, approaching 100 microns, might survive as *drizzle* below very low cloud in near-saturated air. Real rain drops are very much larger, often a millimetre in diameter, and some hundreds of thousand times heavier than cloud droplets. How the one grows into the other is a meteorological problem to which at present only theoretical answers have been suggested.

It was formerly thought that condensation continued around a condensation nucleus until the drop formed was heavy enough to fall out of the cloud. But it has been shown that continuous condensation would not produce large drops but merely a very high concentration of small droplets which have no tendency to fall. The crux of the problem is basically that of the introduction into the cloud of some droplets which are larger than their fellows: if this could happen, the larger, heavier droplets would tend to fall relative to surrounding smaller ones. In their downward path, the large droplets would be bound to collide with a number of small ones and each time this occurred the small ones would coalesce with the large. By this process of capturing moisture, a falling droplet would become progressively larger and heavier until it could leave the cloud as a rain drop. The question thus reduces to what gives certain cloud droplets the size advantage that sets the process going.

Two answers have been suggested. One is by the Norwegian meteorologist Bergeron, according to whom the rain drop begins as an ice crystal. The Bergeron theory is based on the co-existence within the same part of the cloud of both water and ice and to understand this we first look at the structure of a rain cloud with the help of Fig. 3.6. This shows several thermal zones, separated by significant isotherms: the attached

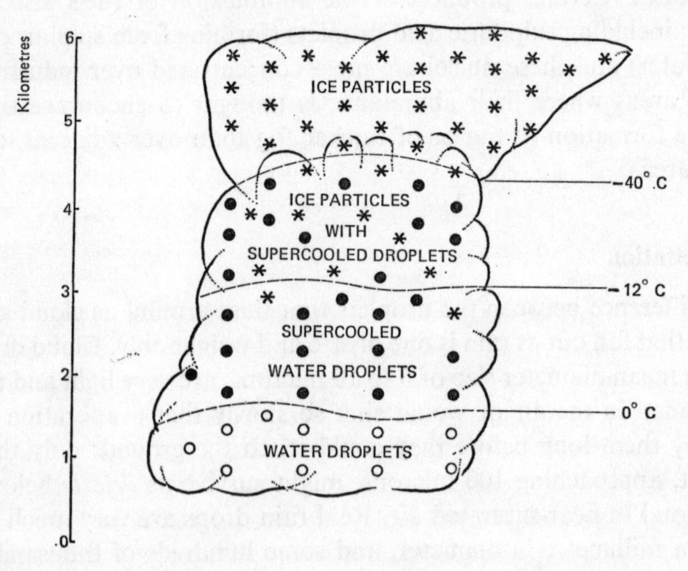

3.6 Structure of a rain cloud.

height scale gives only a crude idea of dimensions, which in fact will vary with the air mass and other circumstances. The lowest and warmest zone, extending from cloud base to the freezing-level (0°C), consists of ordinary water droplets. Above this but below a temperature level of around −12°C (10°F), water droplets are again found but in the *supercooled* condition. It appears that, even in those parts of the cloud that lie above freezing-level, the formation of ice crystals does not readily occur but, like that of water droplets, requires nuclei. These *freezing nuclei* are apparently rather rare in the atmosphere and, although they are thought to be mineral particles with a crystal structure like that of ice, their origin is by no means certain (dust blown from mountain tops, volcanic dust and meteoric dust being among the suggestions). Whatever their source (and it is not even clear whether the ice crystals form from water drops or directly from vapour), the nuclei do not seem to be effective until temperatures as low as −12° to −15°C are reached and only then are the first few ice crystals to be found. With even colder conditions, the ratio of crystals to supercooled droplets steadily increases until temperatures lower than −40°C (−40°F), when freezing occurs readily. The top-most zone thus contains only ice.

The 'Bergeron mechanism' operates in the all-important transitional zone of the cloud where a few ice crystals mingle with many supercooled droplets, both surrounded by air saturated with water vapour. We have explained saturation as a condition of balance in the interchange of water molecules between a water surface (or water droplets) and the adjacent (or enveloping) vapour: a similar equilibrium can be achieved between ice particles and surrounding vapour, but not on quite the same terms. Air that is just saturated in relation to water is super-saturated with respect to ice. Put another way, ice exerts a stronger attraction for the wandering molecules than does water: this is not surprising since ice is a solid, holding newly arrived molecules in a tighter grip and allowing a lesser rate of escape. So when ice and water co-exist in vapour-laden air, water molecules flock from the vapour to the ice crystals: this upsets the equilibrium, leaving the air unsaturated with respect to the liquid droplets, some of which evaporate in order to retrieve the balance. The process continues so that the ice crystals grow rapidly at the expense of the water droplets.

In this way, the crystals eventually become heavy enough to fall, capturing more water in their path, aggregating with other particles to form snowflakes but, once below freezing level, melting to form large rain drops (which can grow further in the normal water zone of the cloud). A further contribution to the theory (by Findeisen) suggests that when water freezes on to an ice crystal, tiny splinters of ice are ejected to form new centres for similar activity elsewhere in the cloud: in this way, there can be a rapid increase in the number of effective crystals.

There is considerable evidence for the feasibility of the Bergeron process as responsible for most of the rain of temperate regions at least. Observations show that rain is most likely from big clouds with sufficient vertical extent, in fact those that reach the ice crystal level. In winter, when this is at lower elevations, precipitation may fall from clouds of modest upward development: on the other hand, summer clouds may reach considerable heights without achieving the necessary level. Similarly, in the cold air of Polar regions, precipitation – which, of course, remains in the snowflake form – can occur from quite thin, low clouds, while great towering clouds in the tropics may yield not a drop of rain. And often enough we learn from the weather forecast that showery weather is expected in northern districts but that 'showers will fall as snow over high ground'.

Not all rain, however, can be explained by this process. There is clear evidence from warm maritime climates that substantial rain often falls from cloud that does not reach the freezing let alone the ice crystal level. For the explanation of this 'warm rain' from 'warm clouds' is needed some alternative mechanism for introducing large droplets which can grow at the expense of their fellows. One such alternative may be the 'Bowen-Ludlam process' (called after the two meteorologists responsible for the suggestion) which depends on coalescence on to extra-large droplets which owe their existence to giant nuclei (probably of salt, reaching the air from breaking wave crests). It is uncertain whether this mechanism is significant in temperate latitudes. Thick clouds are necessary for coalescence – 2000 m (6500 ft) has been suggested as a minimum depth – so that the falling drops can collect many droplets. Probably in temperate regions, with this required cloud depth, the tops are in any case often in the ice-plus-supercooled-water zone, so that the Bergeron mechanism is also available. Despite some uncertainties, both processes are understood sufficiently to have provided the theoretical basis for practical experiments in rain-making (see Chapter 12).

Returning finally to the water cycle from a global viewpoint, we can appreciate that there must be an overall balance between precipitation and evaporation. Considering the problem regionally, however, we find that, for good reasons that will be explored later, evaporation exceeds precipitation in sub-tropical latitudes but that precipitation exceeds evaporation in middle and high latitudes and also in the Equatorial belt. Clearly, the atmospheric motions involved in the general circulation must be such as to transport moisture from the sub-tropics both poleward and equatorward, with the compensating return flow effected by water movement in the ground, rivers and oceans.

4 Air-Mass Properties

When, in everyday conversation, we describe the weather in such homely terms as warm, cold, bright, muggy, wet, thundery and so on, we are often (whether we think so or not) talking about the properties (qualities, characteristics) of the prevailing air mass. Many of the weather elements we observe are associated with particular air masses and a change of weather is often the result of a change of air mass (with the added complication that the boundaries between air masses are in any case zones of disturbed weather). We are not in this chapter concerned with the reasons for air masses but with the properties by which we describe and identify them. An air mass is loosely defined as a large portion of the atmosphere with approximately uniform properties in the horizontal direction: we qualify 'approximately' because the concept is not a precise one and there are good reasons for local and regional variations and 'in the horizontal direction' because, as the last chapter made clear, there are changes in the vertical in all kinds of air (although the particular way in which the properties vary with height is itself characteristic of the air mass).

In our weather experience we are most aware of temperature and moistness. These are in fact the basic or *primary properties* of air masses. It follows from considerations in Chapter 3 that both the temperature and humidity characteristics of the air are acquired from below and it is partly because the opportunities for heat and moisture transfer between surface and air vary over the face of the earth that the atmosphere divides itself up into the distinctive portions we call air masses. We must now show that all the other air-mass characteristics – including cloudiness, visibility, likelihood of precipitation and others less readily defined – depend on the primary properties of temperature and humidity and on their vertical distribution.

Temperature

When we speak of the temperature within an air mass we refer (unless we specify otherwise) to the 'surface' temperature which really means that measured in the Stevenson Screen. Screen temperature, although measured at roughly the height at which we breathe, does not necessarily well represent our subjective impression of warmth or cold. We may feel equally comfortable in an air-conditioned interior regulated to a supposed ideal temperature and in quite cool air out of doors but with the sun shining, and psychologically we may prefer the latter. Thermal comfort is enjoyed when the excess body heat produced as a by-product of our internal chemical processes (*metabolism*) is successfully eliminated by body cooling. This occurs through familiar processes, by conduction and convection losses from the skin to cooler adjacent air, by radiative heat loss to surrounding colder objects and by evaporative cooling due to perspiration. Within a range which for most lightly clothed, inactive people is around 20°C (68°F), body cooling is controlled automatically by the nervous system through the regulation of the blood flow near the skin. At temperatures below the limit of operation of this natural thermostatic control, we feel cold and must check excessive heat loss by wearing warmer clothing or by turning on a heating appliance. At temperatures above the comfort zone, we gain rather than lose heat by radiation and convection, and can lose it only by evaporation, that is by sweating.

Our sensation of warmth or cold is also influenced by humidity and wind strength. Humidity becomes important mainly in warm environments: little evaporation is possible in moist air and, unable to cool sufficiently in spite of continued sweating, we feel hot, sticky and oppressed: in dry air, however, evaporation cooling is more effective and we can tolerate higher temperatures before discomfort sets in. Strong winds are unpleasant with low temperatures as they cause excessive cooling by conduction and convection, but breezes are welcome in hot weather when they encourage evaporation from the perspiring body.

Humidity

The word 'humidity' is a general one signifying the moistness of the air, but it has no precise meteorological meaning. There are, however, several more definite expressions, most of them concerned with the ratio of water vapour to the air as a whole. The water content may be expressed in terms of weight of vapour contained in a unit volume of air (*absolute humidity*), or as the weight of vapour in a unit weight of air (*specific humidity*), or as the weight of vapour per unit weight of dry, i.e. vapourless air (*humidity mixing ratio*). Alternatively, that part of the total pressure of the air which is contributed by its water vapour – the *vapour pressure* – is also a precise measure of the humidity, but expressed in millibars.

For many practical purposes the actual amount of water vapour contained in the air is of less interest than its nearness to saturation. This is given by the *relative humidity*, which is the ratio of the amount of vapour actually present in the air to the amount that could be held at the same temperature: this is expressed as a percentage. Thus, saturated air has a relative humidity of 100 per cent, since the actual amount equals the maximum possible amount, but if the air contains only half the water vapour required to saturate it at that temperature the relative humidity is 50 per cent.

The proviso concerning temperatures is necessary because, as explained in Chapter 3, the vapour-holding capacity of air varies with its temperature. If the temperature is kept constant and more water is evaporated into the air, its relative humidity will increase. If the water vapour content of the air is left unaltered but the temperature is raised, the 'maximum possible amount' is increased and the relative humidity falls: conversely, if the temperature is decreased, the relative humidity rises. This means that, assuming no change in actual water content (such as could happen with a change of air mass), relative humidity must undergo a diurnal variation which reflects the temperature variation but in the opposite sense (Fig. 3.5).

It should be remembered that the same value of relative humidity means different actual water contents at different temperatures. Thus the relative humidity may be 70 per cent on both a winter day and a summer day, but in the latter case the absolute humidity may be twice that in the former. Provided that we do not lose sight of its meaning and limitations, relative humidity is a useful measure of atmospheric dampness in terms of the imminence of saturation.

If we cool a sample of air, without altering its water vapour content, its temperature will fall and its relative humidity will rise, ultimately to 100 per cent when the air becomes saturated. The temperature at which this happens is the *dew-point temperature* (or simply dew-point), which may be defined as the temperature to which air must be cooled for saturation to occur. Any further cooling must result in condensation. Air temperature and dew-point together give another indication of the moistness of the air, for the closer these are, the nearer is the air to saturation, which is one reason why both values are plotted on the weather map.

Relative humidity and dew-point may be evaluated from readings of the dry-bulb and wet-bulb thermometers, found in combination in the Stevenson Screen or as the whirling hygrometer (Chapter 1). The dry-bulb records the ordinary air temperature, the wet-bulb a temperature usually lowered by cooling due to evaporation from the wet muslin. In very dry air considerable evaporation can occur and therefore considerable cooling (as explained in Chapter 3), so that the wet-bulb temperature is depressed far below the dry-bulb. If the air is moist, there can be little evaporation and only a slight depression of the wet-bulb temperature, while if the air is saturated, clearly the two thermometers will read alike. Dry and wet-bulb temperatures, relative humidity and dew-point are all related in hygrometric formulae, from which – by the use of tables, graphs or a special humidity slide-rule – the required values are easily found.

Stability and Instability

If temperature and humidity are the primary air-mass properties, the terms *stability* and *instability* represent a highly important secondary characteristic, deriving from them and their variation with height. These are of course everyday terms and though they have precise meanings in meteorological language, these are not unrelated to their normal usage. Stability and instability refer to the proneness of the air to vertical movement. Stability is a condition in which such movement is discouraged, in which an impetus that might result in vertical motion is resisted, in the way that a stable currency is unaffected by speculation or a lamp-post is indifferent to a push. Instability is the state which favours such motions, which responds to a suitable impetus, in the way that a drunken man, already unsteady on his feet, reacts to a push by falling over.

We can approach this question by first looking back at Fig. 3.3 in which each plotted radio-sounding represents the particular lapse rate in the air mass concerned. These are so-called *Environment Lapse Rates* (ELR), referring to the general distribution of temperature with height as sampled by the ascending radio-sondes and not to be confused with other lapse rates which we shall shortly introduce. The average ELR in the troposphere has already been given as 0·65°C/100 m but a further glance at Fig. 3.3 will remind us that portions of individual

ELRs, particularly in the lower troposphere, may differ substantially from this value.

When air rises in the atmosphere it becomes cooler and conversely when air at some height sinks it becomes warmer. This is not directly connected with the general environmental decrease of temperature with height in the troposphere. It must be remembered that heat conduction in air is negligible and a vertically moving 'parcel' of air may be regarded as fairly effectively insulated from the surrounding still air. The temperature changes with which we are concerned occur because the 'parcel' moves to a level at which it is subjected to a different pressure. We call such a process *adiabatic*.

A descending 'parcel' of air moves into regions of progressively higher pressure (greater weight of overlying air) or, in other words, becomes compressed. In this process energy is expended on the descending air and adds to its internal heat store, so that its temperature rises. The warming up of the lower end of a bicycle pump during the pumping process is a familiar example of warming by compression. When, on the other hand, a 'parcel' of air rises, it moves into a region of lower pressure and therefore expands: in this case, the expanding air itself is doing work against the surrounding air and the energy required for expansion can come only from its internal store, which means a fall in temperature. Cooling gases by expansion has an application in certain types of household refrigerators. We must therefore associate ascent of air with adiabatic cooling and descent with adiabatic warming. The rate at which this cooling or warming occurs depends on the laws governing the behaviour of gases and is constant, assuming the air remains unsaturated. We call it the *Dry Adiabatic Lapse Rate* ('dry' here meaning unsaturated) and it has a value, as near as makes no difference, of 1°C/100 m. For convenience we refer to it as the DALR.

Let us now visualize a small body of unsaturated air near the surface and assume that this is made to rise by some means or other within the larger body of the air mass as a whole. As it rises, it cools (at the DALR) and we must now recall that its relative humidity increases so that it is in fact approaching saturation. Ultimately, with continued ascent and cooling, the air becomes saturated – at the dew-point temperature – and some of its water vapour condenses out as liquid water. But condensation is inevitably accompanied by the release of the latent heat of vaporization (Chapter 3), which has the immediate effect of reducing the rate of cooling, in fact, to about half the dry adiabatic rate. This introduces a new lapse rate – the rate of cooling of rising saturated air (or the rate of warming of descending saturated air) – which is termed the *Saturated Adiabatic Lapse Rate* (SALR).

The SALR differs from the DALR in not being constant. It varies understandably with the moisture availability which in turn depends on the temperature. So while 0·5°C/100 m is a useful figure to remember, the actual value is rather less in the warmest air near the surface and increases with height to approach the DALR in the cold upper troposphere.

We are now in a position to return to the question of stability and instability, with the help of Table 4. Columns 2, 3 and 4 represent the observed change of temperature with height in three quite mythical air masses, *A*, *B* and *C*, in which we suppose uniform ELRs of the

Table 4

Height m	*Column 1 Rising Air DALR 1C°/ 100 m*	*SALR 0·5C°/ 100 m*	*Column 2 ELR 'A' 0·4°C/100 m*	*Column 3 ELR 'B' 1·2°C/100 m*	*Column 4 ELR 'C' 0·8°C/100 m*
3000		−10	−2	−26	−14
2500		−7·5	0	−20	−10
2000		−5	2	−14	−6
1500		−2·5	4	−8	−2
Saturation Level					
1000	0	0	6	−2	2
500	5		8	4	6
0	10		10	10	10

values given and also, for convenience, an identical surface temperature of 10°C. The lapse rates are not necessarily realistic at all heights but they will serve to illustrate the argument. If, in any of these air masses, we apply an upward push to part of the surface air (the impetus referred to on an earlier page), the rising air 'parcel' thus created must cool at the DALR while it remains unsaturated. Let us further assume that humidity conditions in all three air masses are such that saturation occurs at a height of just above 1000 m. Column 1, representing the rising air, therefore shows a DALR in the first 1000 m and an SALR above. This we have taken for convenience, as a uniform

0·5°C/100 m: the departure from truth is not enough to affect the argument. But first we confine our attention to conditions below the saturation level.

The purpose of the Table is to enable us to compare, level for level, the temperature of the environment air with that of an isolated 'parcel' ascending within it. Consider first air mass '*A*' with its small ELR: clearly in this case the rising air (Column 1) rapidly becomes cooler than its environment, the difference being 6° at 1000 m. The *density* (weight of a unit volume) of a gas depends on its temperature: if this is increased the gas becomes less dense (lighter), if it is decreased, the gas becomes denser (heavier). In our example, the rising air is progressively becoming heavier than the surrounding air, which means that it has no inherent tendency to continue rising and, if left to itself, will sink back to the surface: it will rise, as it were unwillingly, only if we persist in pushing it up. This will be recognized as the stable situation.

We repeat the argument, now with the very different ELR of Column 3 and find that in this case the rising air (Column 1) soon becomes warmer than its surroundings, in other words progressively more buoyant: there is nothing to stop it rising with increasing vigour. Air mass '*B*' thus illustrates the unstable situation. Column 4 represents another stable environment, as long as we consider conditions up to 1000 m. It will now be obvious that air (assuming it is unsaturated) is stable if its ELR is less than, and unstable if its ELR is greater than, the DALR. Of course, the ELR may exactly equal the DALR, in which case vertical motion may or may not occur and we describe the condition as one of *neutral equilibrium*.

We now venture above the saturation level, bearing in mind that under natural conditions a rising body of air is usually unsaturated near the ground but is likely to become saturated higher up. Above this level a new criterion for stability must apply since the comparison is now between the ELR and the SALR. But what seems to be a formidable number of alternative conditions can in fact be reduced to three, which adequately define the possibilities of vertical motion. Compare Column 2 with Column 1: despite the halving of the lapse rate in the rising air, the latter remains colder than its surroundings and becomes progressively more so with height. This satisfies the criteria of stability whether the air is saturated or not: we would therefore describe air mass '*A*' as *absolutely stable* and the condition is completely defined as obtaining when the ELR is less than the SALR. It would not alter the argument if air were pushed up from say 1500 m rather than from the surface. In contrast, air mass '*B*' must be described as *absolutely unstable*, since rising air, under these conditions, is always warmer than its environment: the criterion here is that the ELR exceeds the DALR.

A third and interesting alternative, which is of frequent occurrence, is represented by Column 4. In unsaturated conditions we have seen that air mass '*C*' is stable: above the saturation level, however, the situation becomes transformed. At 1500 m the rising air is still a little cooler than its surroundings, at 2000 m it has become 1° warmer and at 3000 m 4° warmer. In short, the air has become unstable. This state is known as *conditional instability* and clearly the criterion is an ELR intermediate between the DALR and SALR. This is instability that lies dormant, as it were, and its realization is conditional upon the air becoming saturated, which in turn depends on sufficient lift to ensure the required amount of cooling.

Temperature–Height Diagrams

In practice the forecaster determines the stability conditions of an air mass by comparing these various lapse rates in terms of their slopes on the tephigram (Chapter 3). The principles may be understood with the aid of the simpler temperature–height graph shown in Fig. 4.1. On such a diagram, any straight line of slope 1°C in 100 m represents a DALR and a number of these so-called *dry adiabats* have been inserted. *Saturated adiabats* (which can be drawn only approximately on this framework) are necessarily curved lines, steeper than the dry adiabats in the lower layers but becoming parallel to them at great heights: however within the 2000 m represented the curvature would be small and it does little offence to the truth to regard the saturated adiabats as straight lines with the slope of 0·5°C/100 m.

The three ELRs of Table 4 have been plotted on Fig. 4.1. We immediately recognize the absolutely stable environment (*A*) as a line of steeper slope than the saturated adiabat drawn from its base: the more vertical the ELR, the more stable is the air, while an ELR actually leaning to the right (temperature inversion) implies a condition of extreme stability. Line *B* represents the absolutely unstable case, leaning more to the left than a dry adiabat drawn from its base, while conditional instability is depicted by *C* which has a slope intermediate between those of the dry and saturated adiabats.

Absolutely stable environments may be found at any level in the troposphere but commonly prevail in the lowest air layers due to the nocturnal cooling of the ground or when warm air moves over colder ground. Absolute instability is practically limited to the air layer near the ground on warm afternoons: the development of convection, which it strongly encourages, tends to reduce the lapse rate higher up to something near the dry adiabatic. As might be expected, given clear calm conditions over land, there is a diurnal variation of lapse

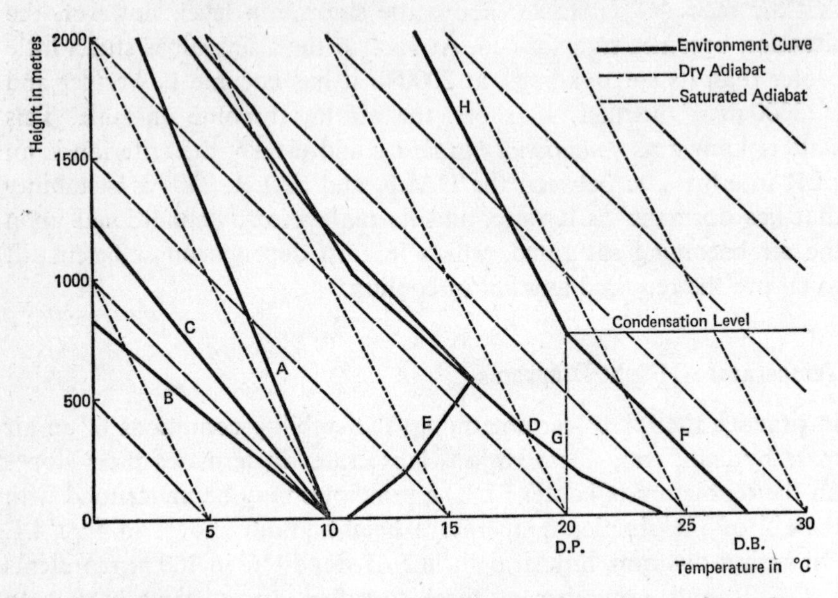

4.1 Illustrating absolute stability (a) absolute instability (b) conditional instability (c) diurnal variation of lapse rate (d, e) and the determination of condensation level (fgh).

rate – and therefore of stability conditions – which is closely associated with the diurnal variation of surface temperature referred to in Chapter 3. Lines *D* and *E* in Fig. 4.1 are based on an actual example: the early morning curve (*E*) shows a marked temperature inversion due to the chilling of the ground and the adjacent air, while later when the sun has become effective, the mid-afternoon lapse rate (*D*) is decidedly unstable in the lower layers. With increasing height above the surface, the effect of the cooling and heating diminishes until a level is reached (about 600 m in this case) at which no diurnal variation is discernible, while above this level, the curves differ little. Mid-morning and evening lapse rates, which tend to be similar, represent intermediate conditions, often with an isothermal layer near the ground. This diurnal change of lapse rate is not to be expected over the open sea.

6. Polar continental weather: rime accretion on the B.B.C. television mast at Holme Moss, Yorkshire. March 1954.

7. Tropical maritime weather: sea fog dispersing over the Antrim coast, Northern Ireland.

8. Tropical maritime weather: a dull stratocumulus sky over Reading.

9. Frontal weather: fine tufted cirrus, often the forerunner of a warm front.

10. Frontal weather: skies clearing over the Brecon Beacons after the passage of a cold front.

11. Thundery weather: towering cumulus and the spreading anvil of a cumulonimbus (thunder-cloud).

The right-hand portion of Fig. 4.1 is used to illustrate a practical use of the temperature–height diagram. Here we depict the behaviour of a small body of air having a dry-bulb temperature of 27·5°C and a dew-point of 20°. If this air is lifted it cools dry-adiabatically and its change of temperature with height is represented by the dry adiabat drawn from 27·5° (*F* on the diagram). Meanwhile the dew-point of this air decreases so slightly with ascent (as long as the air remains unsaturated) that we may fairly ignore this and represent its near-constancy by a vertical line drawn from 20°C (line *G*). Eventually the rising air cools to its dew-point or, in terms of the diagram, lines *F* and *G* meet: the air is now saturated and the height at which this occurs is the *saturation level,* or the *condensation level,* since above it water vapour must begin to condense out to form water droplets or cloud. Above this level (750 m in this case), the now saturated air, if still able to rise, cools at the SALR and its behaviour is depicted by a saturated adiabat (line *H*) drawn from the intersection.

We may now begin to appreciate the value of such a diagram in providing information about the behaviour of air in vertical motion. The line *FH* represents the path of rising air, both before and after saturation and the condensation level is effectively the *cloud base*, under these particular conditions. However, the extent of vertical motion and the possibility of cloud depend, as we shall see, on the stability conditions at all heights, i.e. on the form of the ELR as given by the radio-sounding. It should be admitted that all these considerations involving the temperature–height diagram (or for that matter the tephigram) are just a little approximate. The rising 'parcel' of air is not completely divorced from the surrounding air but in fact incorporates a little of it during ascent (entrainment) and this must affect the rate of cooling. Again, the environment cannot be conceived of as absolutely still: where part of the air rises convectively, the rest must presumably descend (and therefore warm adiabatically). But these effects are of small magnitude and forecasters ignore them in making graphical constructions of the type described.

Up- and Down-Currents

So far we have resorted to a hypothetical 'pushing up' of bodies of air in different environments in order to illustrate the conditions in which vertical motion is likely or not. In nature these up-currents (and the associated down-currents) are produced in various ways and, although we have in Chapter 3 described all vertical movements in general terms as convection, we should distinguish three types by separate names.

The most obvious – and the one that is nearest our hypothetical example – is the forcible lifting of horizontally moving air by high ground in its path: this is *orographic* (or *forced*) *ascent*. Clearly enough, under these conditions, the air must ascend, whether it is stable or unstable, but in the first case, the amount of uplift will be limited, while with initially absolutely or conditionally unstable air the orographic lift will simply set off or reinforce its inherent tendency to upward motion. Orographic ascent, which we consider more fully in Chapter 9, is of great significance for the weather of hilly or mountainous terrain.

On a smaller scale, whenever air moves horizontally over a more or less rough surface it is apt to develop vertical motions. The lowest air layers, continually striking against surface irregularities, are joggled into a series of small-scale circulations or eddies which are carried along with the wind. This frictionally induced vertical movement is known as *turbulence* and it is well illustrated by the progressive diffusion of a smoke plume with increasing distance from a chimney. The eddying, sometimes reinforcing, sometimes countering, the general flow, is responsible for the gusts and lulls we normally experience as wind and which are clearly revealed by the trace of a recording anemometer (Figs. 4.8 and 6.10).

Turbulence is more pronounced the stronger the wind and the rougher the surface: it is generally more effective over land than over sea, though the sea can be rough enough at times. Turbulence is favoured by instability, does occur in stable or neutral conditions but tends to be damped down in the extreme stability of a marked surface inversion. Essentially a mixing process, churning the air as a paddle-wheel churns water, it redistributes heat and moisture throughout a depth of air – the *friction layer* – of between half and one kilometre (say 1500 to 3000 feet): clearly, it plays an important part in the formation and modification of air masses. Within the turbulent layer, because of the vertical motions, the lapse rate tends to become adiabatic.

The third type of vertical motion, *convection* proper, is thermally induced, the result of the heating of surface air by contact with warmer ground: the air, thus warmed, becomes light and buoyant and rises as a convection current or *thermal*. Between the up-currents, air sinks slowly in compensating down-currents so that a circulation is set up rather like that produced by turbulence but generally affecting a much greater depth of air. Convection, unlike orographic ascent and turbulence, is dependent on stability considerations, occurring only in absolutely unstable air, as shown, for example, by the mid-afternoon curve (*D*) in Fig. 4.1. It follows that, over land areas, convectional activity undergoes a diurnal variation, obeying that of stability, reaching its maximum during the afternoon and ceasing at night when the heat supply is cut off.

Convection is most apparent to the observer when it gives rise to cloud, but so-called *dry convection*, implicit of course below cloud base, is rarely evident, except that it occasionally becomes visible on a small scale as the heat shimmer above strongly heated road or other surfaces or on a larger scale in desert and semi-desert regions where dust or sand may be whipped up as *dust-devils* or *sand-whirls*. It is easy to picture convection currents as continuous 'chimneys' of warm air rising above locally over-heated patches of ground but this probably applies only to well organized systems of large dimensions: ordinarily, it is more realistic to visualize them as 'bubbles' of warm air rising periodically from the heat sources (rather as air bubbles rise in water) leaving behind a wake of slightly warmed air which is a favoured zone for the rise of further bubbles.

Pure convection occurs alone only under conditions of calm or very light winds: even with moderate winds, it is inevitably mixed with turbulence in the lower air layers. In fact turbulence may help touch off convection in marginal conditions. In calm air a thermal rises vertically but in even a moderate breeze, its ascent may be at an angle of only 15° from the horizontal. A thermal will ascend to a height dependent on the ELR of the air mass: ascent will cease when the rising air has cooled to the temperature of its surroundings, i.e. when it is no longer buoyant. With a low inversion this will happen quickly and convectional activity is thus limited to the air layer below the inversion. Types of ground surface that warm up rapidly in sunshine are the favoured areas for the development of thermals: these include dark surfaces like some kinds of bare rock or tarmac roads, sand dunes, built-up areas, fields of ripe cereals (but not green vegetation), bare soil sometimes and high land, especially sunny south-facing slopes.

Much has been learned about the nature of convection by careful observations of certain birds (like vultures, buzzards and swallows) and insects (locusts and dragon-flies) that utilize thermal up-currents as an aid to travel. Their ability at thermal soaring has been emulated by glider pilots who have themselves made valuable contributions to our knowledge of the subject. We learn that the speed of up-currents is often 10 m/sec (or 22 m.p.h.) in some large convection clouds and can be much more. Given suitable conditions, experienced pilots have travelled several hundreds of miles using exclusively lift from thermals and have reached upper tropospheric levels in up-currents in cloud. There are however types of soaring other than thermal and pilots also use lift due to high ground.

When large-scale lifting of an entire layer of air occurs, usually at mountain ranges or frontal surfaces, another type of instability must be considered. *Potential* or *convective instability* is illustrated qualitatively in Fig. 4.2. Assume *XY* to represent the ELR of a layer of

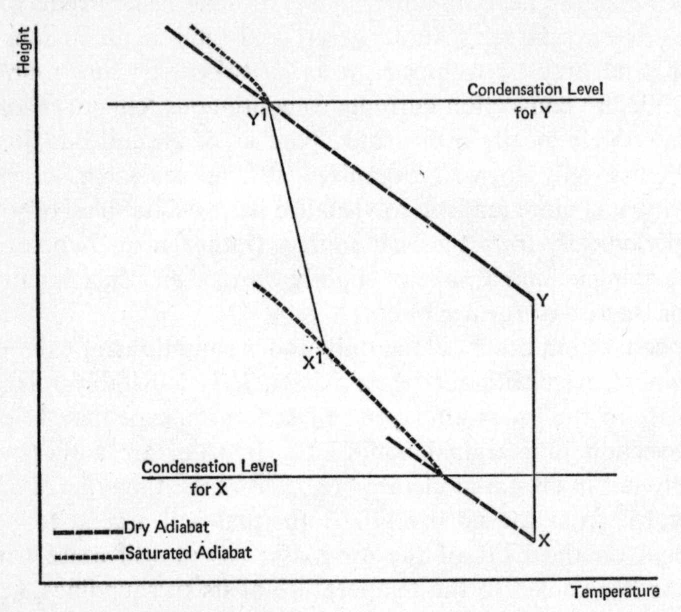

4.2 Illustrating potential instability.

unsaturated air, within which humidity decreases with height. We recognize *XY* as absolutely stable. If the layer is lifted, all the air cools initially at the DALR but the air at *X*, being moister, reaches its condensation level sooner than the air at *Y*. Subsequently, while the air at *Y* is still cooling dry-adiabatically, the air at *X* is cooling more slowly at the SALR. By the time the top of the layer has become saturated, the point *X* has moved to *X′* and the point *Y* to *Y′*: the line *X′Y′* represents the ELR of the air layer after lifting and clearly it is now approaching instability. If the humidity increases with height in the layer, the opposite must occur and the air stabilizes on lifting. But the other vertical distribution of moisture is more common and, in fact, most air masses show potential instability if lifted sufficiently.

Clouds as Air-Mass Characteristics

Clouds result from the cooling of air by uplift to, and beyond, its condensation level. This lifting may occur in various ways, each of which produces typical cloud-forms. We are at present concerned with the processes and resulting cloud that are characteristic of stable and unstable air masses: in the latter case the dominant mechanism is convection, in the former it is turbulence. The clouds associated with uplift at fronts and over high ground are dealt with in Chapters 6 and 9 respectively.

When low-level instability allows convection currents which rise above the condensation level, each is capped by a cloud of the familiar *cumulus* variety (Plate 3). These are often beautiful, sometimes menacing, clouds with flat bases and cauliflower tops. They show up hard and white when lit directly by the sun, dark and with the proverbial silver lining when the sun is behind them. A small cumulus cloud may have a life of only 20 minutes, as anyone with time to watch may confirm, but a large one may be long-lived, with its bubbling, ever-changing tops suggestive of the upsurging currents responsible for its existence. The pattern of clouds in a cumulus sky renders visible the initial pattern of thermals rising from the surface, which again depends on the distribution of heat sources. Sometimes the cloud grouping seems quite haphazard, at others it is surprisingly regular. On occasions the cumulus are neatly arranged in rows – *cloud streets* – and this sometimes signifies the drifting of successive thermals downwind from the same source. With the growth of large cumulus, order in the pattern of the sky is less likely: vigorous convection tends to be self-checking, since the shadows spread over the ground by large clouds may suppress further heating and the development of fresh thermals.

Figures 4.3 and 4.4 illustrate the conditions giving rise to small and large convection clouds respectively. Both are based on radio-sonde ascents made, from the same inland station, at the warmest time of day when convection is most likely. The environment dew-points are plotted as well as temperatures: the distance apart of the two curves gives a ready indication of the moistness of the air.

In Fig. 4.3 the ELR is seen to be absolutely unstable for the first 600 m or so (although except for the lowest 100 m the lapse rate is not far from dry adiabatic, suggesting that convection has been proceeding for some while). The air then becomes conditionally unstable to about 1500 m, above which there is a pronounced inversion with much drier air (as indicated by the separation of environmental tem-

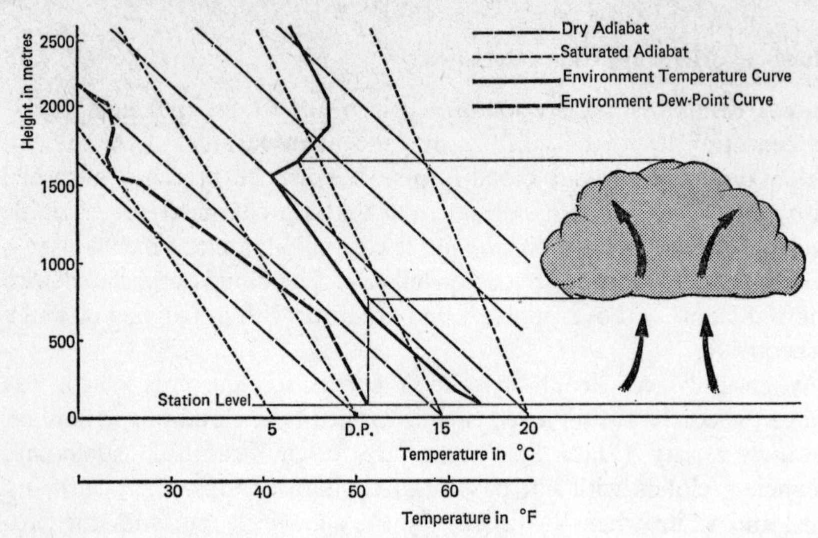

4.3 Fair weather cumulus on the temperature-height diagram.

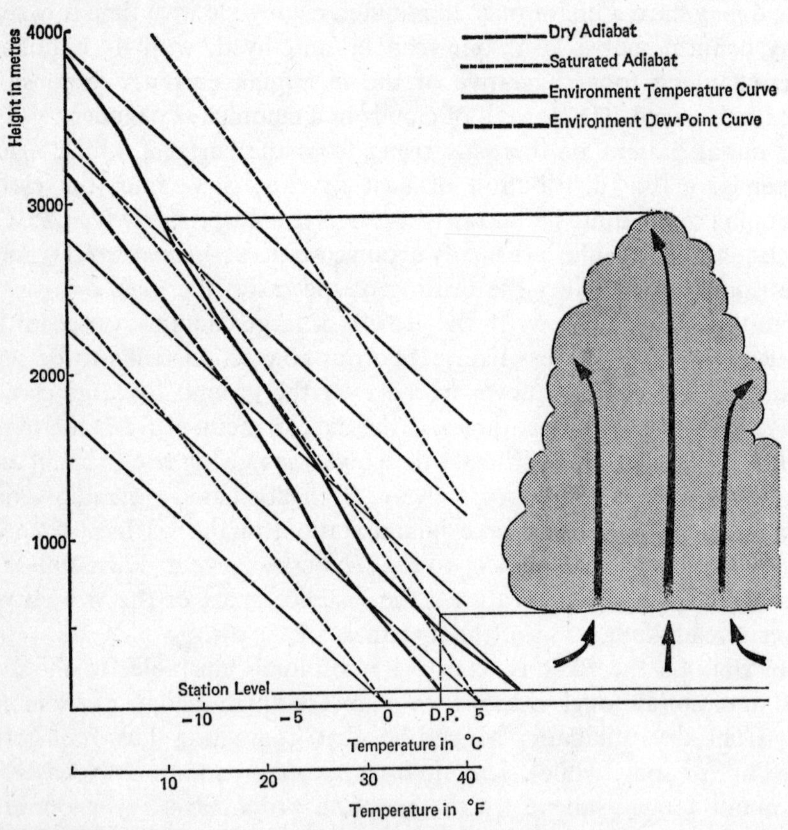

4.4 Large cumulus on the temperature-height diagram.

perature and dew-point curves). The possibilities of convection are given by the simple construction described on page 73. The dry adiabat drawn from the base of the ELR curve and the vertical line drawn from the surface dew-point meet to give a condensation level at about 750 m. From the intersection, a saturated adiabat represents the continued ascent of (now saturated) air: where this line cuts the ELR is approximately the cloud top for here the rising air has cooled to the temperature of the environment and is no longer buoyant. In this example, the presence of the inversion clearly limits the cloud to shallow fair-weather cumulus, with base about 750 m and top about 1500 m (see Plate 13).

In Fig. 4.4, there is no such inversion. Above a super-adiabatic layer of some 400 m thickness the ELR is conditionally unstable to about 900 m and only barely stable above until something like 2500 m is reached. In addition the air is moist at all heights. Clearly the possibilities of cloud are greater in this example. Repeating the construction, we find that a large cumulus, with base about 500 m and tops approaching 3000 m, can be produced. This cloud, unlike the fair-weather cumulus, could possibly yield a shower, as the top just reaches the ice crystal level (see Chapter 3).

The turbulent cloud characteristic of stable air typically has the form of a layer or sheet near the surface. It often occurs under an inversion, for example when warm air moves briskly over a colder surface. The effect of turbulence is to transport heat downwards and to distribute moisture from the surface upwards. The churning motion thus modifies the initial lapse rate and tends to establish adiabatic conditions throughout the layer. If the moisture content of the air is high, the colder air in the upper part of the turbulent layer may well become saturated, so that a low cloud sheet forms, its base perhaps only 100 m above the surface, its top depending on the depth of turbulence.

The process is illustrated in Fig. 4.5 which shows successive ascents at an East Anglian station in a situation in which calm conditions gave way to strengthening easterly winds and low turbulent cloud was forecast for Eastern England. The midnight curve, with a pronounced surface inversion and dry except near the surface, is quite transformed by noon. Turbulence has stirred the air to about 700 m: below this the lapse rate is practically dry adiabatic below and saturated adiabatic above and the layer is now moist throughout. With a surface dew-point of 0°C, a condensation level is possible at about 250 m, the base of a cloud layer some 450 m thick.

The most typical sheet cloud is known as *stratus* but probably the commonest low turbulent cloud is *stratocumulus* (Plate 8). This appears often as a collection of individual cloud masses, quite large in size, sometimes rounded, sometimes elongated in bands or rolls.

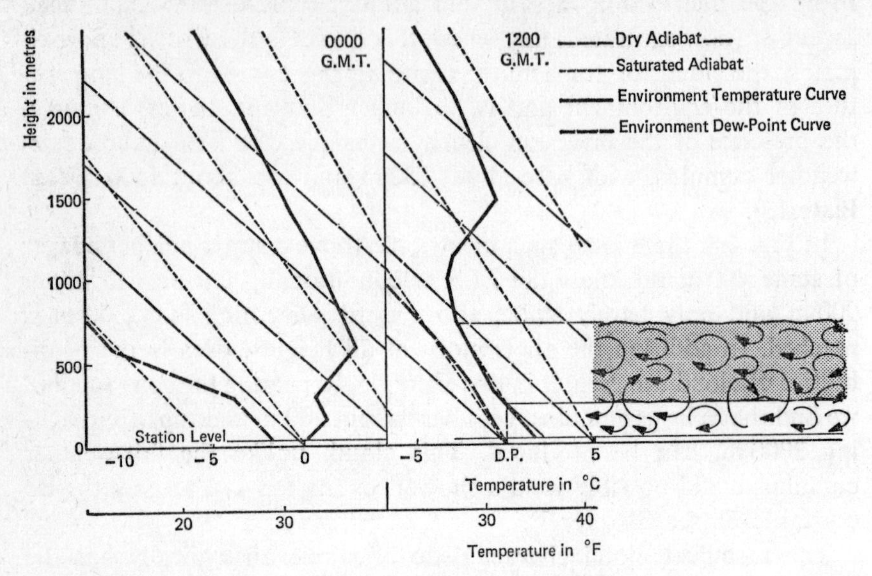

4.5 Turbulent stratocumulus on the temperature-height diagram.

The individual cloud elements may be separated by patches of blue sky (representing the down-currents of warming and drying air in the turbulent motions). With thicker stratocumulus no blue sky may be visible but a definite pattern of light and dark suggests the same structure. Sometimes, however, the cloud sheet appears quite uniformly grey and featureless: this is stratus proper. It resembles fog lifted off the ground and frequently is. Because clouds of this type have only limited vertical extent they do not tend to give precipitation beyond occasional drizzle or light snow (and this is often aided by lift at hills or coasts). More generally they are associated with dry though dull weather.

Cloud Transformations

Nothing is static for long in meteorology. The stability conditions of an air mass or part of it may change and clouds are often transformed from one type to another. There are several forms transitional between layer types and those with vertical development. For example, when convectional activity is limited by a strong inversion the cumulus, being unable to extend higher, spread out laterally to form a stratocumulus layer. The difference between fair weather cumulus and stratocumulus formed convectionally in this way is thus one of degree only. Similarly, at the medium cloud levels, that is from about 2 to 6 km (roughly 7000 to 20 000 feet), the upper parts of large cumulus may spread out into *altocumulus* (the medium level counterpart of stratocumulus).

There is also a general tendency for a stable layer cloud to become unstable, unless it is being cooled from below, for example, by continued movement of the air mass over a colder surface. Otherwise, the base of the cloud warms by constantly absorbing long-wave radiation from the ground, while the upper surface cools by radiation to the colder over-lying atmosphere. In this way the lapse rate within the cloud layer increases and tends towards instability. A stratus sheet thus breaks up into a stratocumulus layer and sheet clouds at higher levels, formed usually during frontal ascent, may similarly be transformed into clouds showing some vertical development: thus *altostratus*, the sheet cloud of medium levels, becomes *altocumulus*, which is similar to stratocumulus except that the individual cloud elements, being higher, seem smaller and *cirrostratus*, the sheet cloud of high levels, composed entirely of ice crystals, becomes *cirrocumulus*, a beautiful and less common type with the appearance of tiny white masses or ripples (sometimes known as 'mackerel sky').

Extreme Stability – Fog

To the meteorologist the term *fog* signifies any atmospheric condition that obscures surface visibility to less than 1 km (1100 yards). With poor visibilities, between 1 and 2 km, *mist* is used if the condition is produced by water droplets and *haze* if it is due to dust or smoke. The man in the street, particularly if he is a motorist, has his own definitions and would probably regard a visibility worse than 200 m or 220 yards as fog ('thick fog' to the meteorologist) because it impedes traffic. We can consider fog as essentially a cloud formed on the surface (while remembering that a smoke pall or even a snowstorm may reduce visibility below the 1 km limit). A simple classification of water fogs separates them initially into those characteristic of stable air masses, which are our present concern, and those associated with fronts (Chapter 6).

Of the three types of air-mass fog, two result from the cooling of air to its dew-point by contact with a cold surface and these are distinguished by the manner in which the cooling occurs. When warm moist air moves over a cold surface, *advection fog* may form. When moist stagnant air is chilled by contact with a surface itself cooled by nocturnal radiation, the result may be *radiation fog*.

Advection fog is formed by the same process as produces turbulent cloud. In the case of a stratus sheet, because of the heat transported downwards by turbulence, the very lowest air layer is too warm for saturation and remains dry below the stratus base. But with less wind and consequently less turbulence, the air remains cold near the surface and there becomes saturated. The formation of this type of fog thus depends on the wind speed: too little means insufficient air movement and too much produces not fog but stratus. In a pioneer investigation of advection fog off Newfoundland (where warm moist air blowing over a cold ocean current in summer provides ideal conditions), it was found that fogs were most frequent with a Force 3 wind (around 4·5 m/sec or 10 m.p.h.).

Advection fog, which is generally about 100 m or so thick, tends to be persistent over the sea but inland, like stratus, it often disperses with day-time warming only to return at night. The clearing process may however go no further than the intermediate stage of 'lifted fog' or low stratus, when high ground remains cloud-capped or shrouded in *hill fog* (depending on the view-point of the observer). An increase in wind speed will also 'lift' the fog, that is, convert it to stratus because of increased turbulence.

Advection fog is largely a sea fog, though it may form over land. Radiation fog however is purely a land fog (though it may later drift over the sea), being the result of nocturnal radiation cooling, which can occur only over land. Clear skies and little air movement (but not calm) are required for the formation of this kind of fog and we shall see that these conditions are typical of anticyclonic situations (see, for example, Fig. 5.4). Figure 4.6 illustrates the appearance of such a fog on the temperature-height diagram: the saturated air (depicted by coincident environment temperature and dew-point curves) corresponds roughly with the surface inversion layer and is about 300 m deep, though most radiation fogs are shallower than this.

The wind speed is again crucial. In the complete absence of wind, the cooling of the air by conduction and radiation alone does no more than deposit dew on the ground or produce a layer of ground fog chest-high at most. For the development of a substantial radiation fog, some stirring is necessary to spread the cooling upwards over a sufficient depth of air. On the other hand, too much turbulence results in stratus rather than fog. Most radiation fogs form with wind speeds of 1 to 3 m/sec (say 2 to 6 m.p.h.). More so than the advection type, radiation fog tends towards localized occurrences: the cleanliness of the atmosphere, the shape of the land and the nature of the surface all influence its formation (Chapter 9).

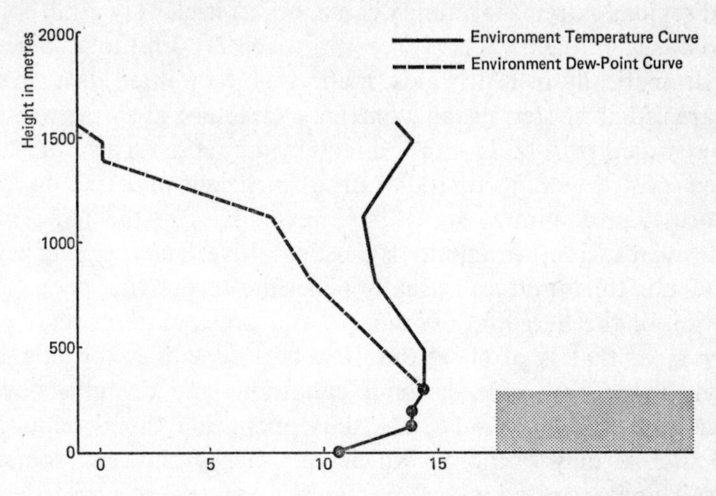

4.6 Radiation fog on the temperature-height diagram.

Radiation fogs occur most frequently during the long nights of autumn and winter. They occasionally persist all day and even several days on end in stagnant situations. At other times of year the sun is sufficiently powerful to penetrate the fog and warm the ground, after which the fog layer is rapidly dried out from below. Summer radiation fogs are fleeting and do not form part of the experience of those of us who rise at a normal hour. An increase in wind speed will also clear, or at least lift, a radiation fog.

The third type of air-mass fog is the result of the transport of cold air over a warm water surface. This places it in the advective category but under these conditions the saturation of the air is due to intense evaporation, not to cooling. A large temperature contrast between water and air seems essential. The cold air is quickly saturated and the ensuing condensation is visible as steaming, hence the name *steam fog* or *steaming fog* or *Arctic Sea Smoke,* since the type is well seen over relatively warm Arctic seas when very cold air blows off adjacent ice- and snow-covered land. It is a fog of little depth, often 15 m, occasionally 30 m (about 50 and 100 feet). On a smaller scale, steaming may occur in cold air above lakes and rivers, from town streets after summer afternoon showers and, of course, in exhaled breath on a cold winter day.

Extreme Instability – Showers and Thunderstorms

In arid regions extreme instability can manifest itself only in dry convection to considerable heights (which can reach 3000 m) and sometimes more dramatically in whirlwinds, made visible by lifted dust or sand: these are rather mysterious phenomena, characterized by their whirling motion (which may be in either direction) about a vertical or slightly inclined axis, extending to 100 m or so in height and usually lasting no more than a minute or two. However, it is in the more humid climates that extreme instability is most assertive. The towering cumuliform clouds thus produced usually penetrate to the vital zone for the operation of the Bergeron mechanism and precipitation occurs of the *shower* type, that is of short duration but sometimes formidable intensity. Under rather exceptional conditions the cumulus develops further into a *cumulonimbus*, the dark menacing thundercloud, and *thunderstorms* may occur, in which very heavy showers, sometimes with hail, are associated with a characteristic pattern of surface weather and above all with more or less spectacular manifestations of atmospheric electricity.

The transition between cumulus and cumulonimbus occurs when the cloud pushes up into the colder regions of the troposphere where water can exist only as ice: ice formation (*glaciation*) therefore occurs in the cloud top, which loses its sharp rounded outlines and becomes fibrous and diffuse. Just as condensation liberates the latent heat of vaporization at lower levels, so glaciation here releases the latent heat of fusion of ice, which supplies another boost of energy for renewed ascent of air. With further development, the top spreads upwards and outwards into a dense mass of *cirrus* (ice cloud) which often assumes a characteristic anvil shape: the flat top is due to the prevention of further ascent by a stable environment (which may be the tropopause) and the cloud spreads laterally with the prevailing wind at that level (Plate 11). The base of a cumulonimbus may be as low as 300 m (say 1000 feet), the tops may reach 9 km (30 000 feet) in this country and 15 km (50 000 feet) in the tropics.

Sometimes the observer may see precipitation trails or *virga* hanging from the base of a distant cumulonimbus: these are falling rain drops or snowflakes that evaporate before reaching the ground. At a later stage the distant shower may show up as a dark curtain between cloud and ground. Sometimes, too, before the onset of precipitation obscures the detail, the underside of the cloud may appear bulbous, hanging in pouch-like protuberances (*mamma*): these may be due to a kind of downward convection touched off by the cooling of the air just below the base because of the evaporation of falling rain, or to the downward drag of heavy droplets (Plate 12).

That thunderclouds contain significant down-currents as well as the expected up-currents had long been suspected but it is largely to a detailed investigation called the Thunderstorm Project which was undertaken during the late 1940s in the United States – and in which surface and aerological stations, aircraft flights and radar were all enlisted – that we owe our present fairly clear picture of the life-cycle of a typical thunderstorm. This involves three distinct stages (Fig. 4.7). In the first or *cumulus* stage, convectional updraughts prevail

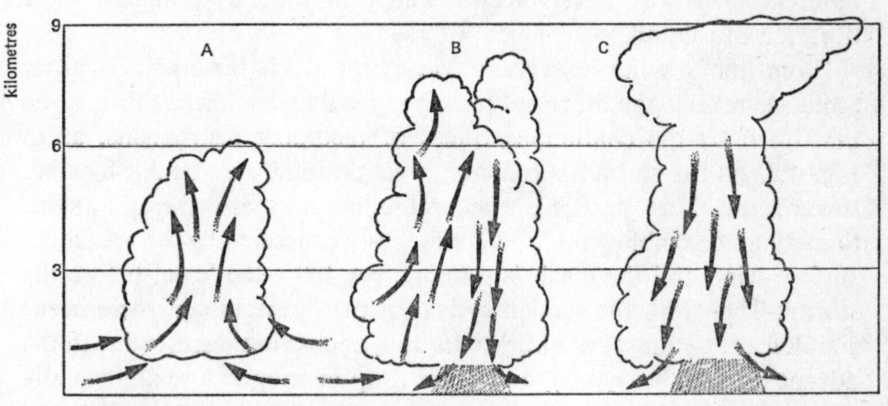

4.7 Life-cycle of a thunder-cloud (a) cumulus stage (b) mature stage (c) dissipating stage.

throughout the cloud, occasionally with speeds exceeding 20 m/sec (45 m.p.h.) though generally less than this. In the surrounding dry air slow compensating downflow completes the circulation or cell. Temperatures are higher within the cloud than at corresponding levels outside, although a large volume of external air is dragged (entrained) in through the sides. The Bergeron process may already be operating but the ice crystals are not yet heavy enough to fall against the updraughts, although they grow steadily as small droplets are swept upwards around them.

Eventually the updraughts can no longer sustain the increasing weight of ice crystals and these fall suddenly as precipitation. In so doing they frictionally drag down cold air from high in the cloud: this is the *downdraught*, which initiates the second or *mature stage*. Precipitation and downdraught air, kept cold by evaporation of rain drops below the cloud base, reach the ground together and the descending current here spreads out as a chilly gusty wind. At this stage there are both updraughts and downdraughts within the cloud, the former now accelerating so that the glaciating tops achieve their maximum height. Downdraught speeds may reach 8 m/sec (18 m.p.h.).

Downward motion gradually extends throughout the cloud and eventually it dominates, the supply of energy is cut off and the cell has reached the final or *dissipating stage*. The precipitation dies away and the cloud dissipates from below, often leaving residual patches of altostratus (which may give light rain for a while) and cirrostratus (the remnant anvil). The whole life-cycle may take an hour but often very large thunderstorms consist of several cells combined, often at different stages of development, which may yield a much longer duration of rainfall.

From the ground observer's viewpoint a characteristic weather sequence reflects the overhead passage of a thunderstorm: this may be illustrated by the continuous traces of recording instruments, as in Fig. 4.8. As the storm approaches, a gentle wind may be felt blowing towards the cloud, part of the general inflow that feeds the updraught. Sometimes this inblowing wind is effectively countered by the prevailing surface wind and the result is a temporary lull – the 'calm before the storm'. Then with the sudden arrival of the downdraught there occur simultaneously the onset of precipitation, a jump in pressure (due to the advent of cold, heavy air), an abrupt drop in temperature and equally sharp rise in relative humidity, strong gusts of wind and a marked change of direction as the downdraught air blows with the storm. The heaviest precipitation comes first, with large drops and an intense rate of fall, perhaps with hail, and this gradually slackens to light rain.

Since the 'Thunderstorm Project' more recent interest in the investigation of severe thunderstorms has centred on the relation between updraught and downdraught. It can easily be shown that the downdraught air can spread many miles from the storm itself, far beyond the edge of the rain area, to give the characteristic gusty wind, pressure rise, temperature drop and humidity increase (Fig. 4.9). But the downdraught would appear to carry with it the destruction of the parent convective system by spreading cold heavy air at the surface and cutting off the warm air inflow. Much evidence however suggests that severe local storms are associated with a marked increase of wind strength

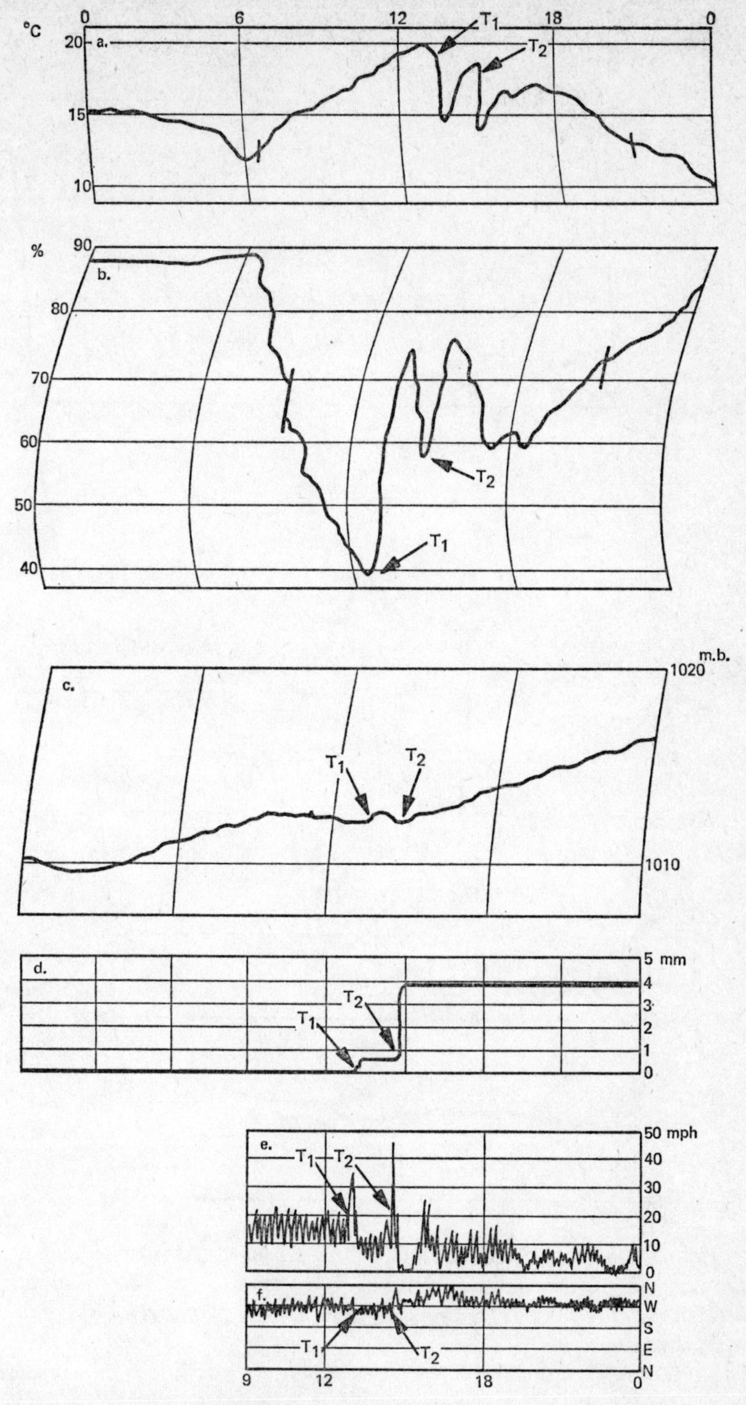

4.8 Thunderstorm weather, illustrated by continuous traces of (a) temperature (b) relative humidity (c) pressure (d) rainfall (e) wind force and (f) direction. Abingdon, Berks., 9 September, 1955. The arrows marked T_1 and T_2 indicate the onset of two thunderstorms.

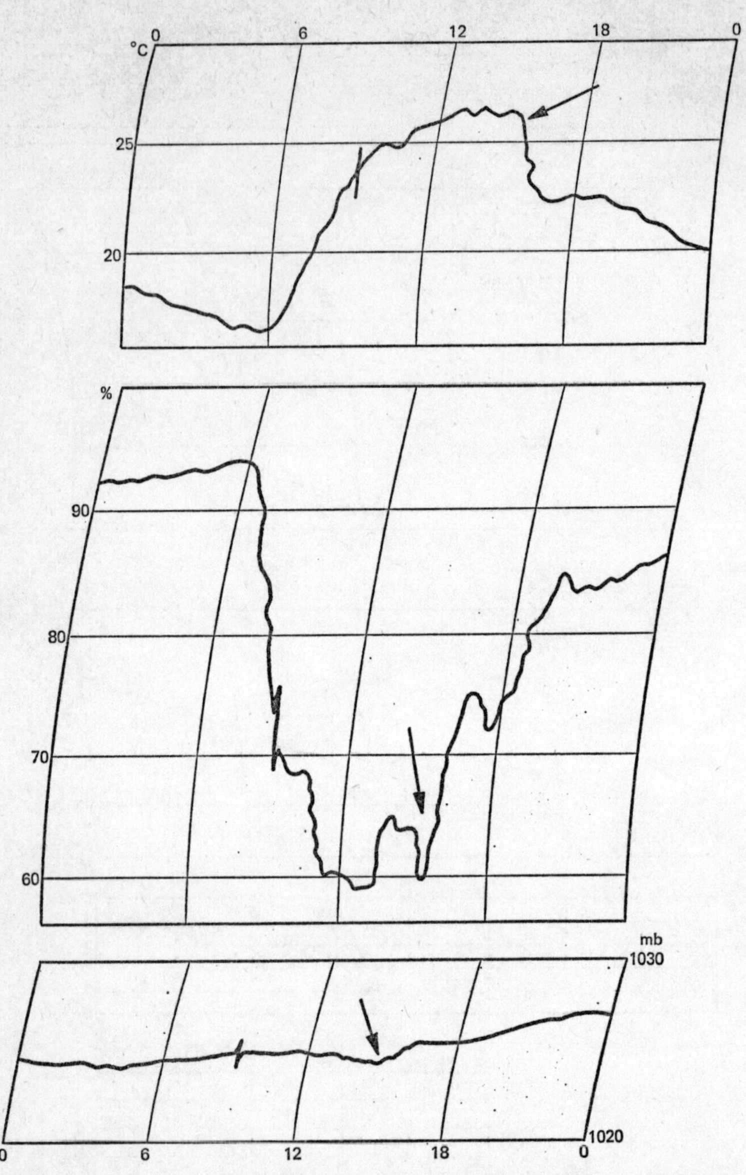

4.9 Thunderstorm weather. The effect of a storm at Tilehurst, Reading, on the recording instruments at Reading University (5 km away) where however no rain fell. 14 July, 1955. Arrows indicate the arrival of cold downdraught air.

with height so that the updraught is itself tilted and the downdraught does not interfere with it. The structure suggested in Fig. 4.10 shows rather that the downdraught air, scooping forward like a shovel, actually re-invigorates the system by forcibly pushing under the in-flowing air in its path. From another point of view, strong winds aloft also help the updraught 'chimney' to 'draw' well and it is usually the powerful upper currents that control the direction of movement of the storm.

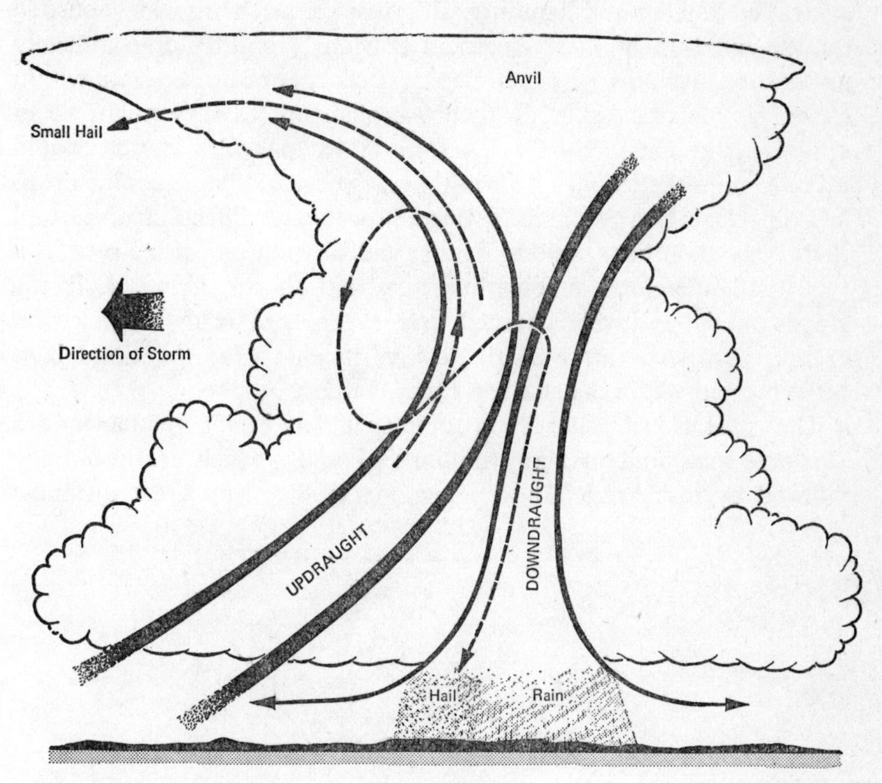

4.10 Suggested structure of a severe storm.

Figure 4.10 also portrays the suggested origin of large hail. Hail-stones are clearly the result of continued accretion of water around an ice crystal, itself the normal product of the Bergeron mechanism. When large water droplets are captured by the ice particle, freezing is slow because of the large amount of latent heat available for release and this gives a layer of clear ice around the original particle: when small droplets freeze on, the process is rapid and much air is trapped, giving a layer of white opaque ice. Thus the onion-like structure of successive layers of opaque and clear ice visible when a large hailstone is cut in half is explained by its passage through parts of the cloud of differing water content. This may well entail several circuits within the cloud before the ever-growing stone finally falls out to reach the ground.

Thunderstorm Electricity

The most dramatic feature of the thunderstorm is the electrical activity evident as thunder and lightning, although the mechanisms responsible for the build-up of these electrical tensions are still not completely understood. It is thought that electrical charges could separate within clouds when large water drops are broken up (as they might be in strong updraughts), by the collision of ice particles (which would occur high in the cloud) or during the splintering of supercooled droplets on freezing. Any or all of these processes could be involved and there may be others besides. The actual distribution of charges in a typical thundercloud is something like that shown in Fig. 4.11: the cloud top is positively charged and everywhere below is negative, except for one or more small positive patches near the cloud base often located where the heaviest rain is falling.

The separation of charges in some such way builds up tremendous tensions, measured in many millions of volts, which are eventually relieved by discharges. These giant sparks that leap great distances

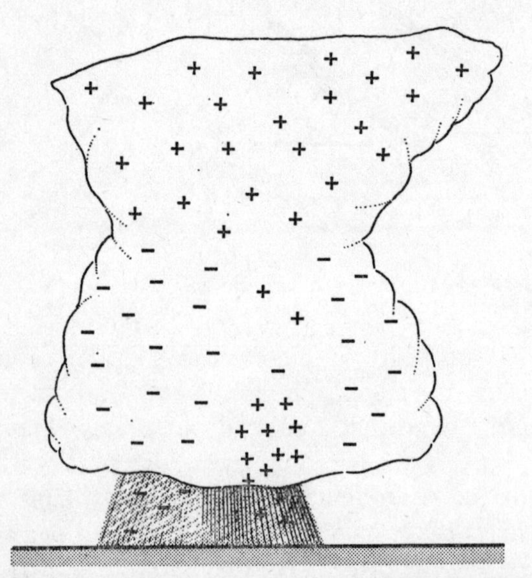

4.11 The distribution of electric charges in a thundercloud.

from cloud to cloud, or cloud to earth, or between different parts of the same cloud, are what we observe as *lightning*. The popular distinction sometimes drawn between 'forked lightning' and 'sheet lightning' is without basis: the latter is simply the momentary illumination of distant clouds by lightning 'forks', themselves hidden by other clouds. A kind of discharge around a central nucleus is occasionally observed and given the name *ball lightning*, but very little is known about this phenomenon. The prodigious heating of the air in the narrow path of the lightning discharge results in a sudden expansion followed by contraction which sets up the sound waves, augmented by echoing effects, that assail our ears as *thunder*. The approximate distance of the thunderstorm is estimated by timing the interval between flash and thunder, counting 3 seconds for a kilometre (or 5 for a mile).

Not all thunderstorms develop within the air mass. Some particularly well organized storms occur in association with fronts (Chapter 6) and the features just described apply just as much to these. The air-mass thunderstorms may result from local over-heating or may be 'triggered off' in various other ways which involve some consideration of local factors (Chapter 9). Some effects of thunderstorms and the problem of forecasting them are dealt with in Chapters 11 and 12.

Conservative and Non-Conservative Properties

It remains for us in this review of air-mass properties to note that while some remain fairly constant with the passage of time, others are very subject to change. Clearly, an absolutely constant property would be a boon to the forecaster, who could use it as a kind of identification label enabling him to trace an air mass easily from chart to chart. For practical purposes no property quite fulfils this requirement, but some are more *conservative* than others.

The dry-bulb temperature is a good example of a non-conservative property: it varies diurnally, it changes adiabatically when the air moves up or down and it is much affected by local factors. Although it is plotted on synoptic charts and provides valuable information of other kinds, it is generally too fickle to act as an air-mass tracer. Among the humidity elements, relative humidity has the same defects.

Of the properties we have considered (there are others that lie beyond the scope of this book), the most conservative is the dew-point, which varies only slightly except when the actual moisture content of the air is altered. There is a diurnal variation if dew is deposited at night, since this must depress the dew-point a little. It is again subject to small local variations: it will increase after a shower (due to evaporation) and is a little higher over coastal areas than inland. But its change with ascent or descent is slow and, in general, the dew-point is a valuable air-mass label, easily determined and plotted on synoptic charts or upper-air diagrams. The secondary characteristics, like cloud type or visibility, are liable to variation for many reasons but often give useful supplementary clues to the identification of air masses.

5 Air-Mass Types

The air-mass concept is basic to the re-thinking of synoptic meteorology by the Norwegian School during the First World War. Now more than 50 years old, the original notions have not escaped criticism, re-examination and amendment. We now think in terms not so much of uniformity of properties as of areas (often of sub-continental dimensions) of 'slack meteorological gradients', compared with the relatively sharp gradients that exist at the air-mass boundaries (fronts). We recognize that the concept has most validity in the lower half of the troposphere and in middle and high latitudes: it becomes blurred at higher tropospheric levels and tends to break down within the tropics. Used with an awareness of its limitations, the concept remains of value not so much as an analytical tool but as a convenient expression of some common units of our everyday weather experience.

Knowledge of the typical air-mass properties, discussed in the last chapter, enables us to condense a lengthy weather description into a few pertinent words. For example, 'cool moist unstable air' adequately conjures up a picture of air of certain origins, convectional activity, towering cumulus clouds and showers alternating with bright intervals: such air also has a characteristic but less definable 'feel' and it is to this total atmospheric environment that we as weather 'consumers' react, rather than to the separate elements of which it is composed. In this chapter we consider the origins of air masses, how they assume the qualities by which we distinguish them and how these may be modified, leading to a classification that will help us understand and describe regional weather, particularly of the British Isles.

Origins and Modification

The essence of the air-mass concept is that the basic properties are impressed on large bodies of air from underlying surfaces. We have already seen that both the temperature and moisture characteristics of air are quite strongly controlled by the surface beneath. The degree of homogeneity that we recognize in air masses depends firstly on their residence over large and fairly uniform areas of the earth's surface. These so-called *source regions* must be entirely sea, or entirely land regions (for example, desert or snow-covered lowland) which exert a fairly uniform influence throughout their extent. It is also necessary for the air to remain over these regions long enough (a week or more) to allow the processes of heat and moisture transference to operate. This means that source regions can be effective only in the relatively stagnant parts of the general circulation and these, we shall see, are the great permanent and semi-permanent high pressure centres, where air descends slowly and spreads outwards, so that external influences are excluded. Fed from the source region, the air moves away, carrying its inherited properties with it to a significant extent, ultimately towards adjacent low pressure areas.

The major source regions of the world, fulfilling both geographical and meteorological requirements, are the polar regions of ice-bound water or snow-covered land, the sub-tropical deserts, the sea areas under the sub-tropical high pressure systems and the great continental interiors when they are in the grip of winter snow. Thus air masses are initially warm or cold, wet or dry, according to the nature of the source region.

These considerations may seem to lead to the commonsense view that, shall we say, today's weather is warm and muggy because it comes from a warm and muggy place, but this rather over-simplifies the situation. In fact, as air moves away from its source region the properties initially inherited are significantly modified by passage over surfaces of different character. Consider for example an air mass formed over the North Atlantic in the sub-tropical high pressure centre well known to us as the 'Azores High': this is a warm moist air mass originating over water of surface temperature 15–21°C (60–70°F). Some of this air habitually spreads northwards, thus passing over progressively colder seas: the air mass is then cooled from below and its initial stability strengthened, with all that this implies for its associated weather. On the other hand, air of the same origin also moves equatorwards, eventually to regions where the sea temperature is over 27°C (80°F): this air is warmed from below, its initial stability is eroded and it becomes an unstable airstream associated with a very different type of weather. Thus the loss or gain of heat from underlying surfaces is an important air-mass modification.

Equally significant is the loss or gain of moisture. A land-derived air mass is dry and therefore thirsty: once it moves over water it may greedily take up moisture. On the other hand, an air mass born over the sea is initially moist: if it moves over land, it can only lose moisture by precipitation but this process requires help from other factors and in their absence the air may remain quite humid after deep penetration into the heart of a continent.

It will be noticed that when an air mass stabilizes on account of its movement, both the surface cooling and further acquisition of moisture are restricted to the lowest (turbulent) layer, usually under a temperature inversion. But when an air mass becomes unstable by virtue of its movement, both the heat and moisture gained are distributed upwards to greater heights by convection currents.

Another kind of air-mass modification is related not to surface influences but to large-scale vertical movement within the air mass or part of it, which are imposed by changes in the associated pressure patterns. The modifications take the form of general uplift (lessening stability) or general sinking (increasing stability). These dynamic effects are best understood in connection with the depressions and anticyclones with which air masses become involved (Chapter 7).

We may appreciate that 'old' air masses that have wandered far from their source regions may eventually become modified out of all recognition so that the immediate path of the air becomes more important than the original source. Again, if an air mass moves over a region of diverse surface properties (for example, part land, part sea), different portions become modified in different ways and the notion of near-uniformity within the air mass becomes invalid. These are obstacles to an uncritical acceptance of air-mass notions and terminology.

Air-Mass Classification

This cautionary note sounding in our ears, we turn to a classification of air masses based on the two considerations outlined above – source region and subsequent modification – which is essentially that devised by J. Bjerknes and his colleagues of the Norwegian School. Theirs was essentially a middle-latitude view-point: cold sources (with exceptions) lay to poleward and warm sources to equatorward, hence an immediate division into *Polar* (P) and *Tropical* (T) air masses. Cutting across this temperature division was a further distinction dependent on whether the source was land or sea: thus air might be *Maritime* (m) or *Con-*

tinental (c), which tells us something about its initial endowment with moisture. There are then four combinations based on the source region – Polar Maritime (Pm), Polar Continental (Pc), Tropical Maritime (Tm) and Tropical Continental (Tc).

A suffix is sometimes added to indicate modification by surface heating or cooling. An air mass is labelled W (warm) if its lower layers are warmer than the underlying surface: this means incidentally that it is absolutely stable at its base. A cold air mass (K for kalt) is colder in its lowest layers than the underlying surface and has a tendency to instability. If the air is moving along rather than across the isotherms representing the prevailing temperature distribution, then neither W nor K can apply. It must be emphasized that these suffixes refer purely to the temperature contrast between air and adjacent surface. From the 'Azores High', Tm air spreads both poleward as TmW and equatorward as TmK, but TmK is eventually warmer than TmW. On the other hand, Pc air is always PcK, since it originates over the coldest regions on earth.

Other categories are sometimes added to the basic classification. For example, *Equatorial Air* refers to air of varied origins that has moved to and become stagnant over warm equatorial waters: it is not an air mass historically speaking but the result of continual 'tropicalization' in the sense of acquisition of warmth and moisture. Such air may also become involved in monsoon circulations (see Chapter 8) when it is often called *Equatorial Monsoon* (Em) air. The term *Mediterranean Air* is sometimes used for air of European origin much modified by long residence over the Mediterranean Sea. The basic classification is universally applicable only because it is simple and generalized. Once we look at particular regions we find that all sorts of modifications become necessary in an attempt to move nearer the realities of the situation.

Air Masses of the British Isles

The air masses that directly affect our islands are conveniently classified in Table 5, which also summarizes very briefly some of the details that follow. The more usual routes by which these air masses reach the British Isles are shown in Fig. 5.1. Some of the major types are illustrated by synoptic charts (Figs. 5.3 to 5.7) chosen to show their characteristic weather patterns. These weather maps use the style of the Daily Weather Report maps (see Chapter 2), with the addition

Table 5 Air Masses of the British Isles

Air Mass	*Source Region*	*Path*	*Direction of Approach*	*Occurrence*	*Typical Cloud*	*Typical Weather*
ARCTIC A (Direct Polar maritime) (PmK)	Arctic Ocean	Ocean Short Track	N, NE	All seasons, not common	Cu, Cb	Cold, showery, sometimes thundery
POLAR MARITIME Pm (PmK)	Arctic Ocean	Ocean Long Track	NW, W	All seasons, common	Cu, Cb	Cool or cold: showers and bright intervals: radiation fog
POLAR MARITIME RETURNING Pmr (PmW)	As for Pm	Ocean Very long track	S, SW	All seasons	St, Sc	As for Tm but cooler
POLAR CONTINENTAL Pc (PcK)	Eurasian Continental Interior	(1) Mainly land	E, SE	Winter	Little cloud small Cu	Very cold and dry
		(2) Land and North Sea	E, NE	Not common	Often St, or Cu	Cold and cloudy: sometimes like Pm
TROPICAL MARITIME Tm (TmW)	North Atlantic Azores region	Ocean	SW, S	All seasons, common	St, Sc	Warm, humid, cloudy: advection fog
TROPICAL CONTINENTAL Tc (TcW)	North Africa	Mainly land or over Med. Sea	S, SE	Summer, not common	Perhaps Cu, Cb	Hot, dry, dusty: sometimes thundery

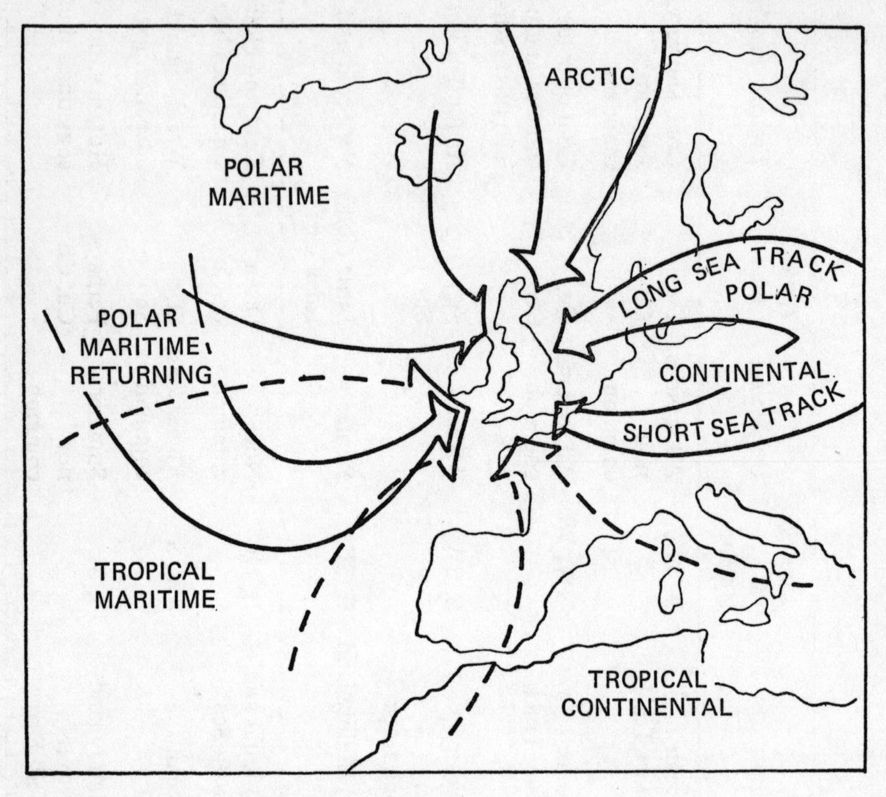

5.1 Air masses affecting the British Isles.

of the station pressures (these will show how the isobars, which for the moment we can accept as broadly depicting the surface air flow, have been drawn in). The charts are all for 0600 hr GMT, that is, they illustrate conditions before day-time heating becomes effective. Typical (but not necessarily average) upper air conditions are shown in Fig. 5.2: these are early morning ascents illustrating nocturnal conditions, the subsequent day-time modification of lapse rate near the ground being suggested by the dotted curves.

The most detailed study of the characteristics and occurrence of air masses over the British Isles was made by J. E. Belasco of the Meteorological Office, and we have generalized his data in Table 6, which gives some idea of the surface temperatures to be expected in each type in south-east England. These are average figures about which a variation of a few degrees is likely, especially when account is taken of local peculiarities (see Chapter 9).

Polar Maritime Air

Much of the air we experience originates in the weak high-pressure area of the Arctic Ocean and its land fringes. If such air spends a

Table 6 Average Maximum and Minimum Temperatures (°C) at Kew in Different Air Masses

	J	*F*	*M*	*A*	*M*	*J*	*J*	*A*	*S*	*O*	*N*	*D*
A	2	4	6	9	12	16	18	17	14	8	6	3
	−2	−2	0	3	4	8	11	10	7	2	0	−2
Pm	7	8	10	13	16	19	21	21	18	13	10	8
	2	2	4	6	8	11	13	12	11	7	4	3
Pmr	10	11	12	15	18	21	22	21	21	15	12	11
	7	6	7	8	11	13	15	14	13	9	8	7
Pc	−1	1	4									2
	−4	−3	−1									−2
Tm	11	12	14	15	19	21	22	22	21	17	14	12
	8	8	9	10	12	14	16	16	14	11	9	8
Tc	9	10	15	20	24	26	28	28	23	18	14	11
	4	4	7	9	13	15	16	16	14	11	9	7
All Air Masses	7	7	9	13	17	20	22	21	18	13	9	7
	2	2	2	4	7	11	13	12	10	6	4	2

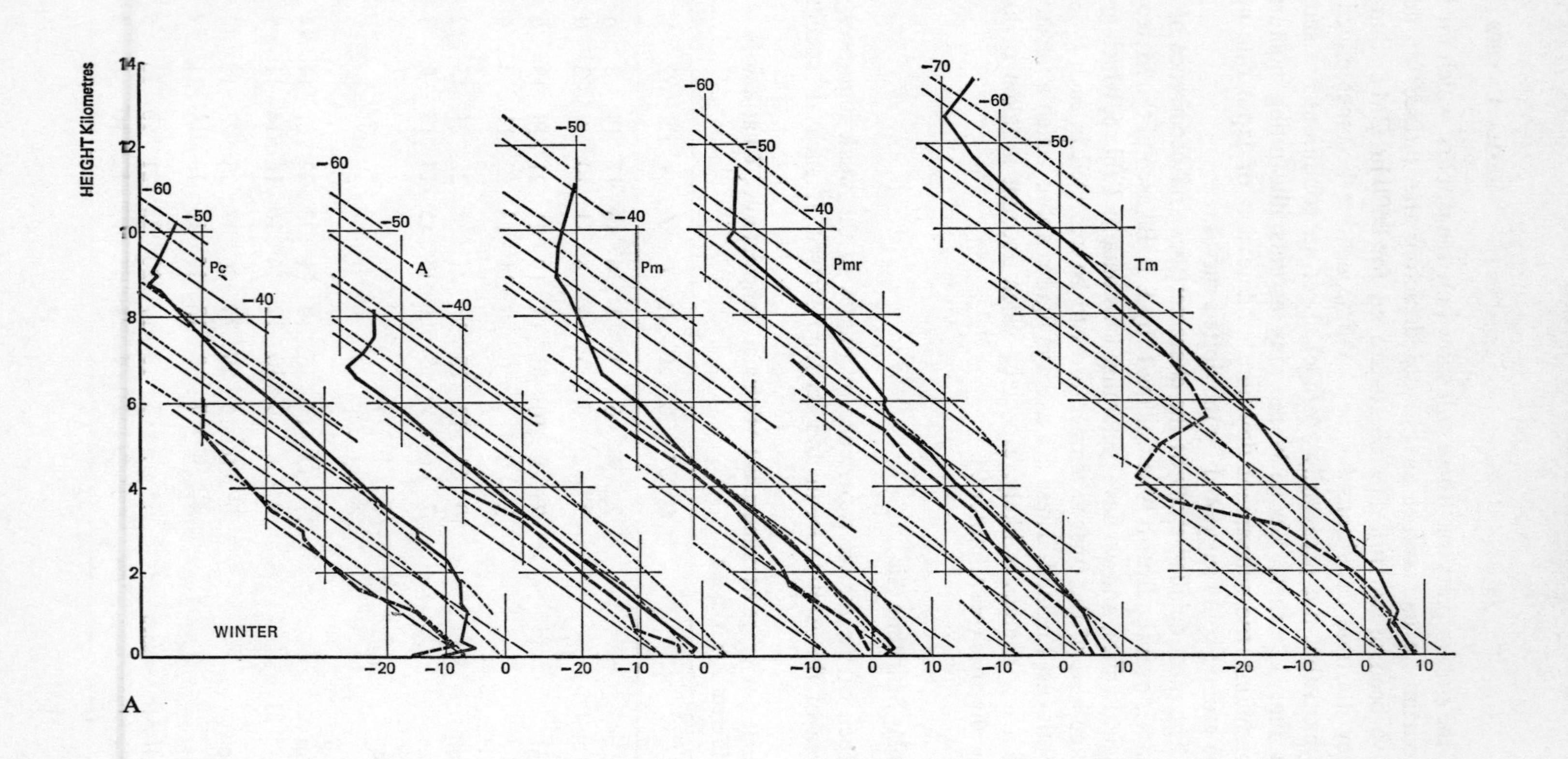
HEIGHT Kilometres
Pc
A
Pm
Pmr
Tm
WINTER
A

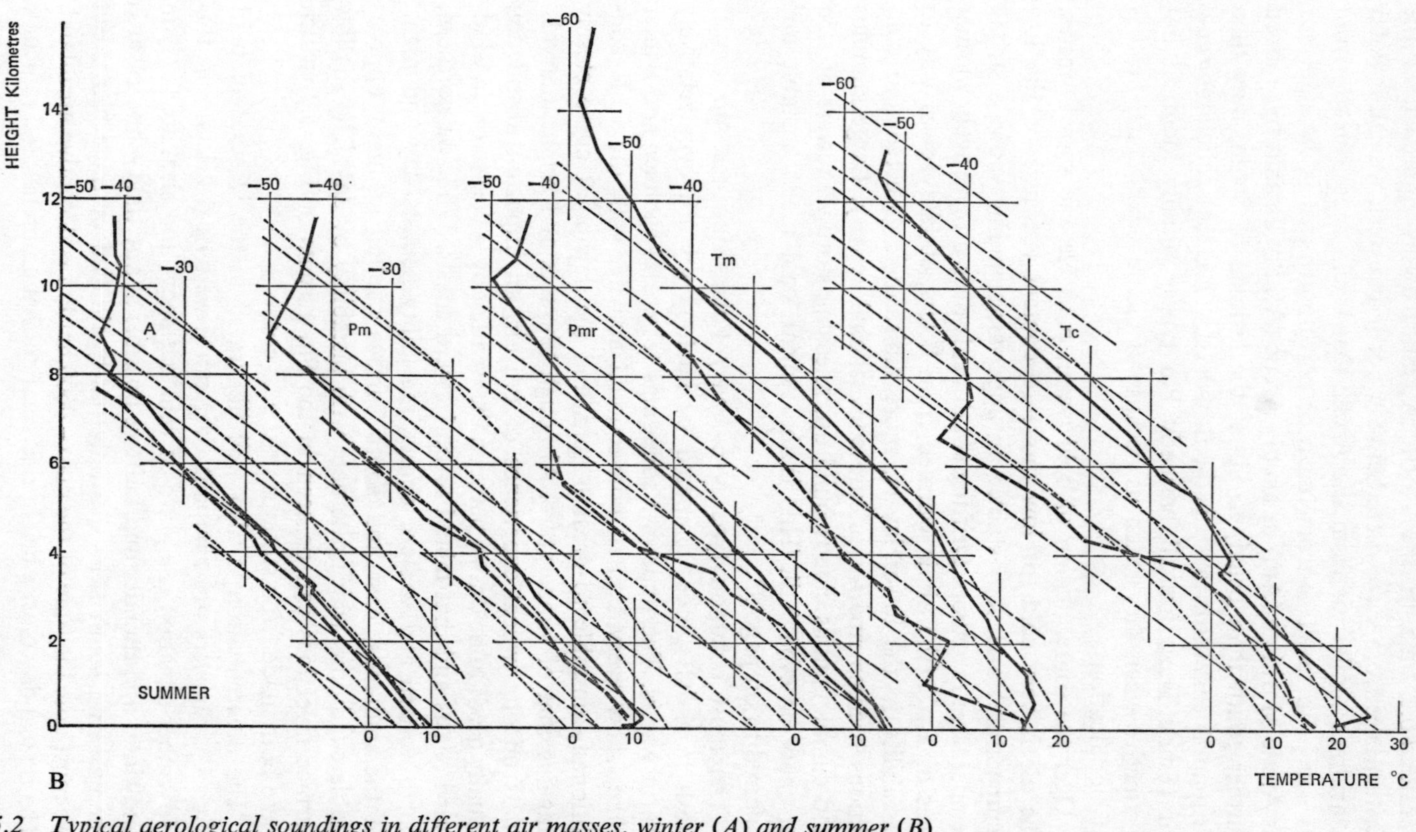

5.2 *Typical aerological soundings in different air masses, winter (A) and summer (B).*

considerable time in the North Atlantic, finally approaching the British Isles from somewhere between north and west, we call it Polar Maritime: similar air from the Arctic that tracks more directly from its source to reach us from north or north-north-east is distinguished as Arctic Maritime (Am) or simply Arctic (A). Pm air that has moved further south than usual, to about the latitude of Spain, may then turn north-eastwards towards the British Isles: it is then called *returning* air (Pmr). These are all basically Pm types, initially identical but distinctive when they reach us because of different paths and degrees of modification.

Over the source region of cold water, pack ice and snowy wastes, the air is very cold, of low moisture content (because of the low temperature) and very stable: an upper-air sounding shows a strong temperature inversion. Drifting southwards over progressively warmer seas as PmK, the air is warmed from below, stability gives way to instability in the lower layers, more moisture is gathered and conditions become ripe for intense convectional activity. Large cumulus and cumulonimbus clouds build up, releasing stormy showers of rain or snow or even hail. This is the typical 'cool moist unstable air' referred to on page 93.

The most common type is ordinary Polar Maritime, no longer fresh from its source, tempered by its long sojourn over relatively warm waters. The upper-air soundings (Fig. 5.2) summer and winter alike, show every sign of convectional activity, with the ELR conditionally unstable to great heights and the dew-point curve closely hugging that of temperature. Although 'showers and bright intervals' sums up our typical experience of Pm air, an important distinction usually needs to be drawn between the weather over the sea or on windward coasts and that inland or on leeward coasts. While on northern, north-western and western coasts, instability persists day and night, in the midlands, east and south surface cooling at night imposes stable conditions which alternate with instability by day. The shallow surface inversions can be seen in the Pm ascents of Fig. 5.2, which were both made from stations near east coasts.

The differentiation is also well illustrated in the weather map of Fig. 5.3. Showers are rampant over north-western districts – in this November situation, they are of rain but snow and sleet are likely in midwinter and thunder and hail are possible at any time – but towards the east and south skies are only partly clouded or quite clear. But we must note the time of the map – 0600 hr! Pm is deceptive air inland. The typical day dawns bright and clear so that umbrellas and raincoats are left behind and the washing is hung unwisely on the garden line. Some time during the morning the cumulus clouds appear, small and innocent at first, large and threatening later (Plate III). The showers usually fall at the least opportune moment and emphasize the lesson

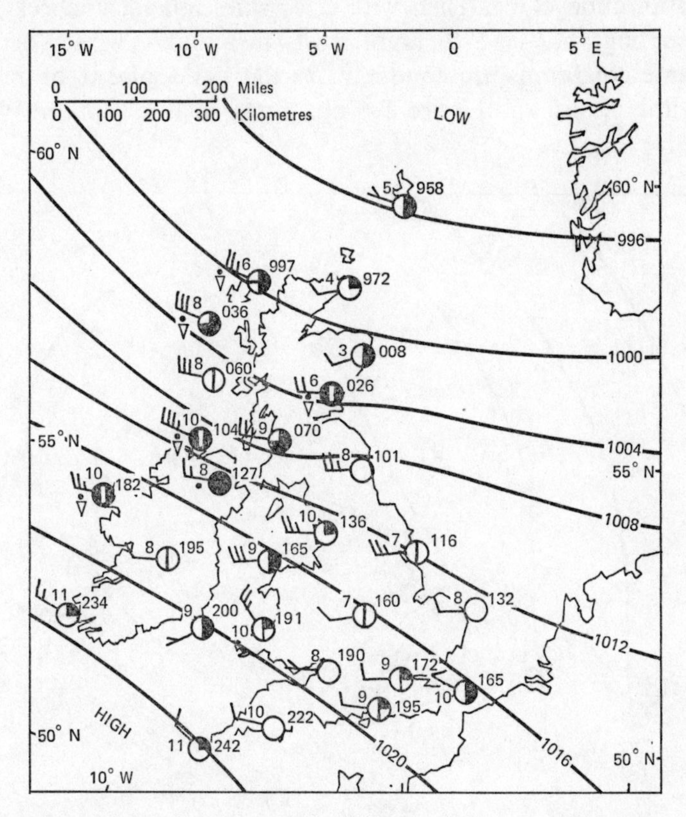

5.3 Synoptic chart. Polar maritime air, autumn (*5 November, 1970, 0600 hr. GMT*).

to be learned (Plate IV). In the evening the showers die down and the clouds disperse to give a fine starlit night.

Yet Pm weather is generally quite pleasant. In between showers, the sun shines brightly and a rainbow is often seen against the departing rain-cloud adding colour and interest to the sky. In winter the air is cold and sometimes there is frost at night, in summer it is on the cool side and often bracing. With the constant stirring of the air by convection and turbulence, dust and smoke are diffused upwards and the atmosphere has a sparkling clarity, though the visibility may deteriorate rapidly enough in the heavier showers. Given warm clothing and some protection against rain, Pm provides good brisk walking weather even in winter and quite reasonable conditions for outdoor sporting spectacles.

On occasion however, Pm air becomes relatively stagnant in the grip of an anticyclone. Then with clear skies inland, unchecked radiative cooling may lead to night frosts in autumn, winter or spring. The same conditions are conducive to the development of radiation fog in this moist air. Figure 5.4 illustrates such a situation with an

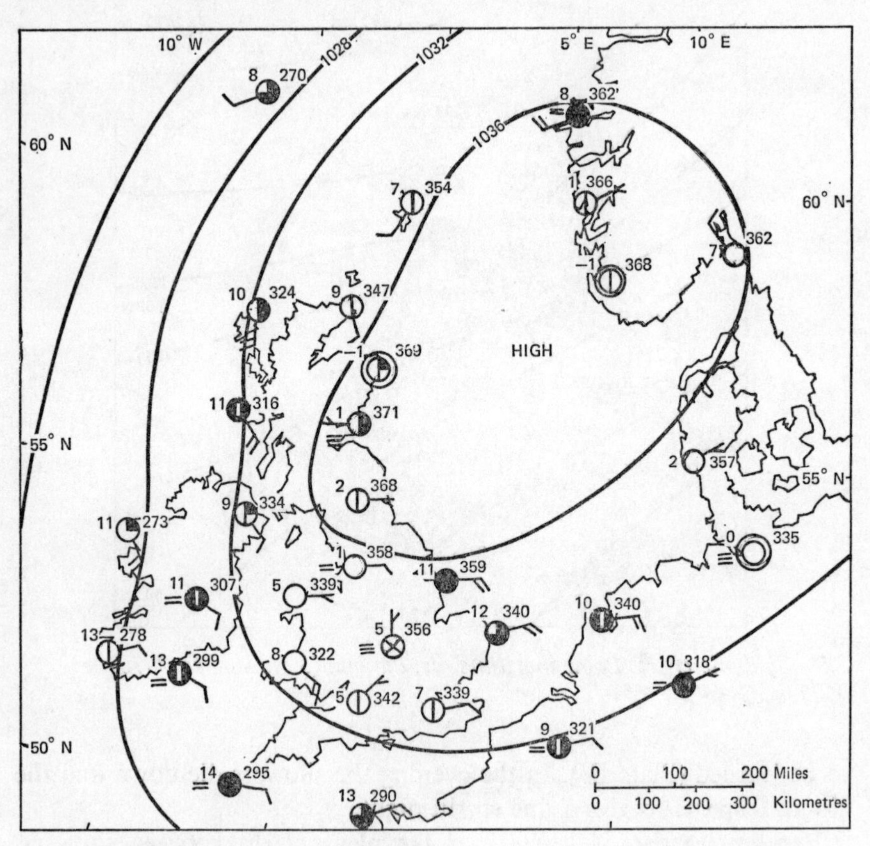

5.4 Synoptic chart: Polar maritime air, autumn anticyclone (26 October, 1971, 0600 hr GMT).

anticyclone centred over the North Sea extending its influence over midland and northern areas. Near-freezing temperatures in the stagnant central part of the high contrast with more normal conditions for October on the fringes of the system: the fog and mist symbols can also be seen. This is the worst of Pm air, when surface transport may be seriously hampered if not totally paralysed and aircraft grounded, while persistent fog, especially when aggravated by pollution and combined with cold, is physiologically distressing to many people and may be lethal to some (see Chapter 10). From the meteorological point of view, such a development illustrates a transformation from the unstable to the stable state.

Arctic air is effectively a fresher, colder variant of Pm. The upper-air soundings (Fig. 5.2) are similar but lower down on the temperature scale. A typical synoptic situation (Fig. 5.5) may be compared with the corresponding chart for Pm (Fig. 5.3): the time of year is almost identical but the Arctic air, with its ominous northerly track, is markedly

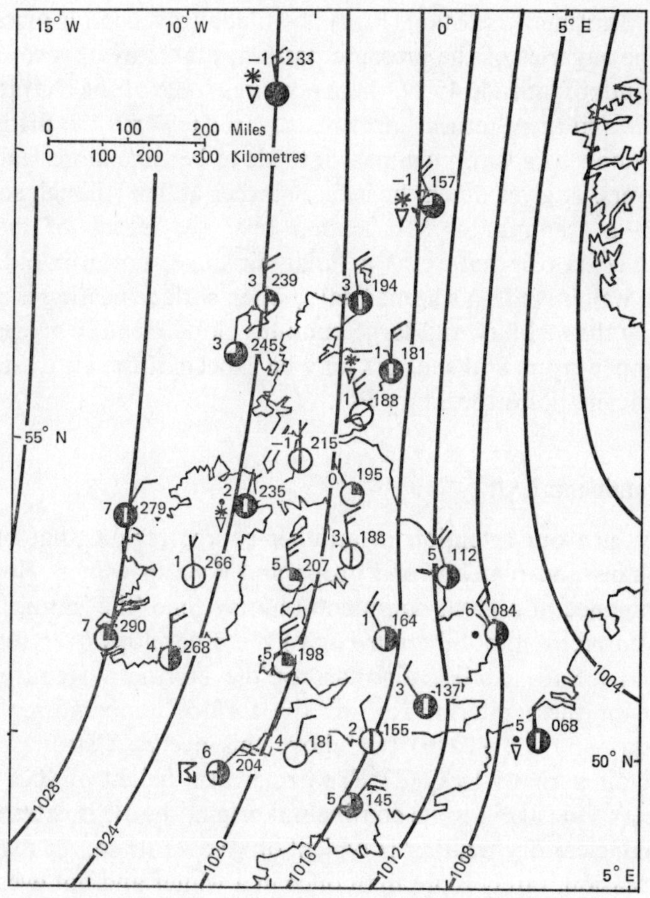

5.5 Synoptic chart: Arctic air, autumn (9 November, 1971, 0600 hr GMT).

colder. The showers, which tend to affect mainly northern and eastern coastal districts, are of snow or sleet rather than rain. A thunderstorm reported at Scilly reminds us that this is very unstable air. During the day showers are everywhere prevalent.

Although not the coldest air we experience (Polar Continental has that distinction) Arctic air, in mid-winter, is probably the least pleasant. High relative humidity combines with cold to give a raw chill in the air and often strong gusty winds and wintry showers make outdoor conditions bitter and harassing. Frost is to be expected at night, especially in the north. In summer, Arctic air provides somewhat cooler interludes: it is quite common in the north, much less so in the south where with rather low maximum temperatures, but less cloud and less frequent showers than in winter, it gives disappointing rather than unpleasant weather.

Polar Maritime Returning (Pmr) is a much travelled air mass which, due to the vagaries of the pressure pattern, moves over warm Atlantic waters south of latitude 45°N, there acquires some of the characteristics of Tm air and then 'returns' north-eastwards towards the British Isles. Behaving now as a warm air mass it tends to become more stable in its lower layers. It gives much the same weather as Tm (though somewhat cooler) – a triumph of modification over the legacy of the source region. Yet its true nature as a Polar air mass, however old, can be betrayed well inland on summer days, when surface heating is sufficient to destroy the stability and large cumulus clouds readily grow into the colder upper layers and showers may be expected: the air then resumes the characteristics of ordinary Pm.

Polar Continental Air

We now turn our attention to another source region, that of the interior plains and plateaus of Europe and Asia in winter. Here under the dominance of a great continental anticyclone (the 'Eurasian High' or, less correctly, the 'Siberian High') there develops over surfaces of snow and ice the coldest air mass of the Northern Hemisphere. In Pc air over these interior regions the surface temperatures may be as low as −30°C (−22°F) but above this occurs a strong inversion with the air perhaps 15°C (27°F) warmer at a height of 1000 m or so (3000 feet). Due again to its continental origin, the air is extremely dry. Very cold, clear dry weather is typical of Pc over its source region.

On occasion, rarely more than once in a winter and not every winter at that, the Eurasian High sends an icy tongue westward and Pc air may then invade the British Isles. A glance at the aerological ascent (Fig. 5.2) will show the extent to which the original properties may be retained. The surface air is no longer as cold as it was at its source but it is cold enough and the inversion aloft is still well maintained. Although some moisture has been acquired in the very lowest layers, little stirring is possible under these very stable conditions and the air above is dry. With day-time warming, a little shallow cumulus cloud might form beneath the inversion but often skies are perfectly clear night and day.

The weather brought by Pc air depends largely on the track it follows and the degree of modification it undergoes. The above description applies to air that has approached from east or south-east, over a mainly land track, crossing only the English Channel or the narrowest part of the North Sea: this relatively little modified Pc is sometimes described as the 'short sea-track' type (see Fig. 5.1). A less pure variety sometimes reaches us from rather north of east, having crossed a wide stretch of the North Sea where water temperatures approach 4° or 5°C (around 40°F) over large areas even at the coldest time of year. Considerable acquisition of warmth and moisture must occur with this 'long sea-track' type. Often turbulence and convection combine to produce a layer of low stratus cloud beneath the inversion: this cloud sheet may spread inland into eastern districts. Sometimes the warming process may entirely eliminate the inversion, in which case the air is transformed into a kind of Pm bringing higher temperatures generally and convection cloud and snow showers to east coast areas.

The synoptic chart of Fig. 5.6 catches a moment during a brief space when Pc air swept in from somewhere over Russia on the southern flank of an elongated ridge. The most characteristic temperatures are over neighbouring continental countries but it is also very cold for March over the British Isles. Cloud is rare except where the air has followed a long sea track. Pc is not necessarily unpleasant, even with sub-freezing temperatures night and day such as may occur with the short sea-track variety: given suitable outdoor clothing, many people find this clear, dry, crisp weather stimulating, especially when it is associated with traditional (though rare) mid-winter activities (Plate V). The less cold but dull weather of long sea-track Pc is often less palatable and the fresh winds afflicting the east coast in Fig. 5.6 would, with these temperatures, drive most people indoors.

Ice deposits are often part of the scene in Pc air. Air frost (i.e. a screen temperature of less than 0°C) may occur in any Polar air mass but actual deposition of ice crystals (white frost or *hoar-frost*) requires also sufficiently moist air. If, as with a little-modified Pc type, the air is very dry, the temperature may never fall low enough to reach its dew-point (or, more properly, *frost-point*) although it may be well below 0°C. The result is then a statistical frost that leaves no sign on the ground (a so-called *black frost*). In moister Pc air under freezing

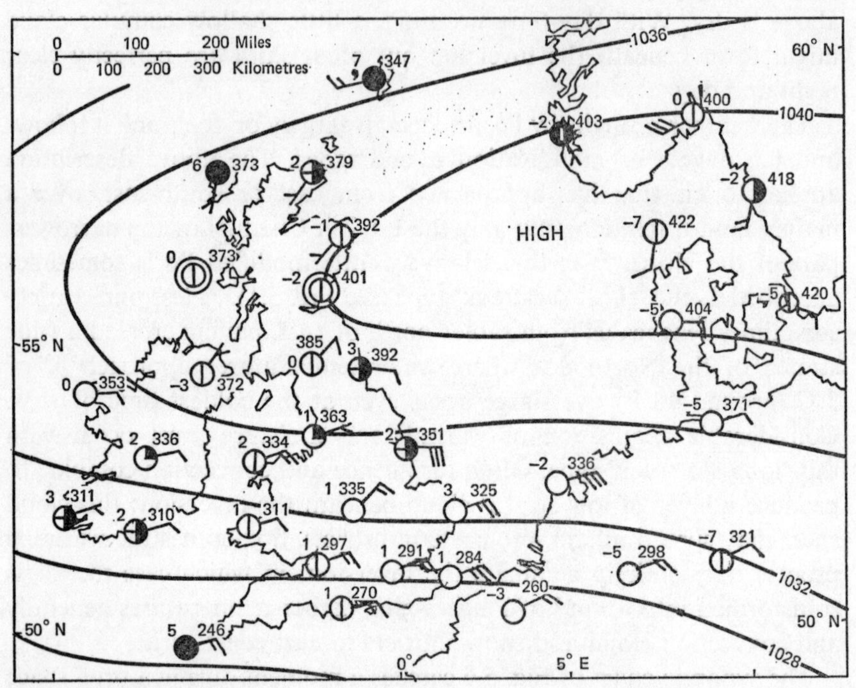

5.6 Synoptic chart: Polar continental air, early spring (*12 March, 1972, 0600 hr GMT*).

conditions a supercooled water fog may form. The droplets tend to freeze on contact with objects like trees and fences, covering them with a coating of opaque ice (*rime*), which helps to create a photogenic winter landscape. With a persistent supercooled fog, rime may grow out in spiky protrusions on the windward side (Plate VI).

Less picturesque is '*glazed frost*' or '*glaze*', which may occur when a long residence of Pc air has left surfaces well below 0°C: if rain then falls (e.g. from a fast advancing warm front) it freezes instantaneously on meeting the ground to give a treacherous coating of smooth clear ice. Glazed frost is rare in lowland areas of Britain but is periodically reported in upland country where considerable damage to trees and overhead wires may be caused by the heavy burden of ice.

Even at its best, Polar Continental air tends to outstay its welcome after a very few days. The great Pc invasions of 1947 and 1963 were reminders that we are not well adapted to the sort of winter weather that is normal to Warsaw or Moscow. January and February of 1963 combined to give the coldest two-month period recorded in central England since 1740, parts of Scotland registered temperatures below −20°C (−4°F) and the water froze in several harbours as well as in many rivers.

True Pc is a winter air mass for only then are the seasonal anticyclones established over the continental interiors. Air that reaches us by easterly routes at other seasons is sometimes called Pc but is really Arctic or Pm much-modified. In summer, air of diverse origins may warm up considerably over Europe and reach us from the east or south-east as a warm air mass akin to Tropical Continental and sometimes so called. This is an instance when air-mass terminology becomes confusing.

Tropical Maritime Air

We have already referred to this air mass without naming it when illustrating the potency of modification (page 94). Reaching us from a warm water source under the 'Azores High', the air is properly designated TmW and is stable in the lower layers, often with a surface inversion. Because of their warmth these layers can acquire a great deal of moisture and the total water content is on average higher than in any other air mass: but the stability discourages upward transport of moisture and the middle and upper layers are relatively dry, as is best shown by the summer ascent in Fig. 5.2.

Sea (advection) fog or low stratus cloud (dependent on the wind speed) characteristically form in this stable air as its bottom layers are chilled to saturation off our south-western coasts. The fog banks rolling in towards these shores may disappoint early holiday visitors, for the frequency of sea fog is highest in late spring and early summer (when the contrast between air and sea temperatures is greatest). The summer upper-air example, an ascent from a Cornish station, shows saturated air in the lowest layers, suggesting a fog sheet about 200 m thick (see Plate 7).

What happens to Tm air on crossing the coast depends on the season, the time of day and the form of the land. In winter, especially at night, the land is colder than the sea and the low cloud persists and may even thicken. Advection fog may spread inland, especially over very cold surfaces, but more often it is 'lifted' by increased turbulence into low stratus. Fig. 5.7 shows a typical winter pattern, with mist or fog on windward coasts and much cloud inland though with breaks evident in the east. High ground tends to be shrouded in 'hill fog' but the scale and amount of detail plotted on the map do not reveal this,

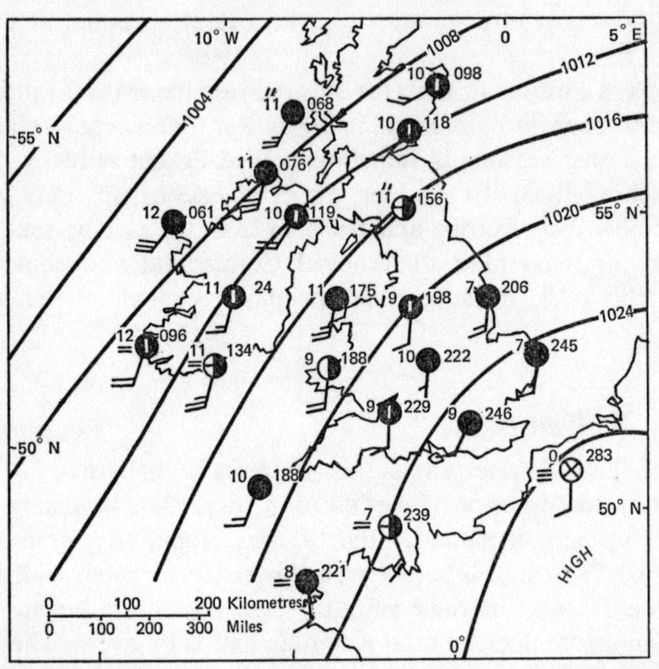

5.7 Synoptic chart: Tropical maritime air, winter (18 December, 1971, 0600 hr GMT).

nor the fact that cloud base usually lifts well inland and the form here becomes stratocumulus (Plate 8) rather than stratus. Precipitation, if any, is usually no more than orographic drizzle or light rain induced by cliff coasts or high ground inland.

Tm air brings our milder winter weather: the temperatures in Fig. 5.7 are generally around 10°C (50°F) even at this morning hour, though not much diurnal temperature variation can be expected under the prevalent unbroken cloud sheet. The high relative humidity however gives the air a mugginess that is less welcome. Many people find this air enervating and sigh for the freshness of Pm at its best. Some claim that aching corns, rheumatic pains and 'twinges' of various kinds accompany – or even presage – the change from Polar air to Tm (though it must be said that the 'hard' evidence available is confusing).

Houses without central or background heating also react to such a change, as the arrival of warm moist air in contact with previously chilled walls gives rise to copious condensation.

The warmth of summer radically alters the potentialities of Tm weather, for daytime heating is sufficient to eliminate the 'W' nature of the air mass everywhere away from the coastal strips. This means that summer sea fog is rarely able to penetrate more than a mile or so inland during the day, while further inland the low cloud soon becomes well broken and may disperse entirely. Further insolation may transform the initial stability of the lower layers into instability and convection cloud can then form. This may be no more than fair-weather cumulus but there is ample moisture for considerable cloud development and continued heating could give rise to strong convection and even to air-mass thunderstorms. In the evening convection cloud disperses and, if night cooling is sufficient, the turbulence cloud may well re-form. Much depends on the age and track of the air: the reversion to stratus is less likely well inland where nights may be fine and clear.

Tm days are pleasantly warm but the air can be oppressive at night when, with temperatures often over 16°C (60°F), sleep may be difficult for some. The combination of warmth and high humidity creates difficulties of various kinds. Bacteria are highly active, milk goes sour and foodstuffs rapidly deteriorate. On the farm, pests thrive and fungal diseases quickly spread. For example, a period of 48 hours or more with temperatures always above 10°C and relative humidities above 75 per cent is liable to be followed two to three weeks later by an outbreak of potato blight. At such times the farmer is busy spraying or dusting with insecticides and fungicides. At its best, however, Tm provides the kind of weather we always demand of our summers and only rarely obtain. When anticyclonic conditions establish themselves in this kind of air we enjoy long spells of sun and warmth – the 'heat waves' of popular parlance – with maximum temperatures soaring to 25°C and above (nearing 80°F) and the sales of ice-cream and cold drinks mounting accordingly. When such fine Tm spells coincide with Whitsun or August holiday weekends, the call to the country or coast is irresistible and the queues of cars outside our major resorts often intolerable.

Visibility is generally poor in Tm air and landscapes take on a hazy, less sharply defined appearance. This is understandable in moist air, which usually needs little inducement to give mist or fog of either advection or radiation type. Otherwise Tm gives generally dry weather, apart from the possibility of convectional thundery rains in summer under much modified conditions. Investigations suggest that some exceptionally heavy falls (mainly in summer) must be attributed to forced uplift of very moist Tropical air over high ground.

Tropical Continental Air

On the morning of 1 July 1968 most people in England and Wales south of a line from Tees-side to Merseyside awoke to find a yellow dust covering car roofs, window-sills and other outdoor horizontal surfaces. Investigation showed that this dust had been raised some three days earlier during sandstorms in the southern Sahara, transported from there in a high-level southerly flow and then deposited in the thunderstorm rain that fell over that part of Britain during the night of 30 June. This was a convincing demonstration that North African air can spread to these islands, even if only at high elevations. In fact we do occasionally experience true Tropical Continental surface air from this source.

In the desert itself and its fringes (and Mediterranean holiday-makers become aware of this) the air is extremely hot and dry and often dust-laden due to strong day-time convection. Occasionally, a southerly flow moves this air towards Britain over a mainly land track: then we scorch under arid skies made pallid by the characteristic dust haze. Or the air may reach us via a more easterly route, spending much time over the Mediterranean Sea, where considerable moisture is picked up and under these conditions (illustrated by the upper-air sounding in Fig. 5.2) Tc may contribute to summer thunderstorms in southern England. Because of its origin, Tc is warmer in its lower layers than Tm and gives us our hottest summer days with maxima near or over 30°C (86°F) and nights, with minima of 16°C (60°F) or more, which may be too warm for comfort. Authentic Tc reaches us mainly in summer but some rare unseasonably warm days in autumn have been attributed to this air mass.

Frequency of Different Air Masses over the British Isles

Some idea of how often and at what times of year these air masses and their characteristic weather are likely to visit us may be gathered from Table 7 which is simplified from data compiled by Belasco. The Table gives, month by month, the average percentage frequency of occurrence of each air mass at three widely separated stations, Kew, Scilly and Stornoway (Outer Hebrides). A comparison of the figures for these stations, situated approximately at the south-eastern, south-western and northern corners respectively of Britain, shows that quite significant differences in the period of occupation by different air masses may be found even in a small country and here is one reason for regional differences of weather and climate within the British Isles. Readers living between the far-flung points represented in the Table can make appropriate interpolation for their own areas.

Table 7 Percentage Frequency of Different Air Masses at Kew, Scilly and Stornoway

Air	*J*	*F*	*M*	*A*	*M*	*J*	*J*	*A*	*S*	*O*	*N*	*D*	*Year*
KEW													
A	7	6	7	7	14	6	5	5	6	4	6	5	6·5
Pm	20	24	16	25	17	32	33	31	25	21	28	24	24·7
Pmr	14	10	7	8	14	6	10	8	8	10	10	14	10·0
Pc	7	6	1	0	0	0	0	0	0	0	0	3	1·4
Tm	11	13	12	6	6	9	11	7	7	8	12	12	9·5
Tc	3	4	5	5	6	3	4	5	3	9	7	2	4·7
F	12	8	10	13	11	11	12	13	13	9	12	11	11·3
H	15	19	31	31	19	30	23	24	35	25	17	22	24·3
SCILLY													
A	4	3	3	4	11	4	3	6	4	2	5	1	4·2
Pm	29	15	18	28	25	34	38	34	25	26	29	29	27·5
Pmr	16	6	12	9	9	5	8	10	11	13	9	11	10·0
Pc	3	6	0	0	0	0	0	0	0	0	0	2	0·9
Tm	17	20	13	9	8	11	15	11	9	12	19	18	13·5
Tc	2	2	4	3	2	1	1	2	3	5	6	1	2·7
F	9	12	11	11	15	13	13	13	12	10	12	11	11·8
H	11	15	30	30	20	29	20	19	34	21	14	22	22·1
STORNOWAY													
A	13	8	10	13	16	14	12	16	9	7	12	5	11·3
Pm	31	30	27	31	23	32	35	31	41	31	31	35	31·5
Pmr	22	20	12	11	15	14	15	14	15	16	16	22	16·0
Pc	1	5	1	0	0	0	0	0	0	0	0	1	0·7
Tm	6	10	13	11	5	6	6	6	7	11	13	10	8·7
Tc	0	1	2	2	1	0	1	0	1	3	3	1	1·3
F	10	9	10	11	15	14	20	14	8	9	12	10	11·8
H	11	11	15	16	19	18	10	15	15	17	7	12	13·8

Belasco derived his frequency values from 12 years of weather maps and they may be taken as a useful guide. Not all the air reaching the British Isles could be classified and the percentages in any one column do not quite add up to 100: most of this indeterminate air came from the Continent without, however, fitting into either of the well-defined Continental types. To the familiar air-mass classes have been added two further categories. F denotes air in the vicinity of fronts, that is, essentially transitional conditions between air masses: H represents anticyclonic air, that is, in or near the central region of an anticyclone where the weather picture owes less to the origin of the air than to the dynamics of this pressure system. Discussion of fronts and anticyclones belongs to later chapters but in passing we may note from Table 7 that the frequency (though not necessarily the intensity) of frontal weather seems practically the same throughout the British Isles, but that anticyclonic weather is much less frequent in the extreme north than in the south.

The Tables show clearly enough that our weather is dominated by maritime air of one kind or another. Even excluding that involved in anticyclones, maritime air is with us for between one-half and two-thirds of the year, the frequency increasing westwards and (especially) northwards. The Atlantic dominates too, providing our air for about half the time (rather less than this at Kew, rather more at Stornoway). The notion of our 'prevailing westerly winds' is another way of expressing this.

Well-defined continental air is relatively rare. Kew and the south-east generally are most likely, for obvious reasons, to experience both the cold Pc of winter and the hot Tc of summer, but even here a year may pass without a visitation from either. Continental air is less common at Scilly and hardly ever reaches Stornoway. This helps to justify the commonly accepted distinction between the more maritime west and the more (but still not very) continental east.

Of the maritime air masses, polar is more important than tropical and Pm is the most frequent of all. If Table 6 is re-examined, it will be seen that the average temperatures for all air masses are closer to those in Pm air than in any other. If we combine Pm and A as giving cool or cold unstable weather, then these persist for about one-third of the time in the south (Kew and Scilly) but for over 40 per cent of the time at Stornoway, when understandably Arctic air is particularly important. If Pmr air is added to Tm, both bringing mild or warm stable conditions, the differences among the three stations are small, though at Stornoway (which has the highest combined quota, nearly 25 per cent), Pmr is nearly twice as frequent as Tm and there are years when no Tm occurs at all.

While generally speaking, Pm and A are more common in the summer than in the winter six months and the reverse is true of Tm and Pmr, each of these is an all-seasons air mass and only Pc is really confined to one particular season. This air is a rare occurrence in December, more common in January and February and is even apt to provide a last snarl of winter in March, as Fig. 5.6 will show.

6 Fronts and Frontal Disturbances

When the air mass in occupation moves away to be replaced by another, we experience the transition from the properties of the first air mass to those of the second. This boundary zone of relatively sharp meteorological (especially temperature) gradient, is a *front* in modern meteorological language and has an overall width of up to 1000 km (say 600 miles) compared with usually many thousands of km extent of the air mass on either side. But more is involved than a transition in air-mass properties. The term 'front' used in this way dates from the First World War and the implied military analogy is no coincidence. Fronts are aptly regarded as the battle lines of the atmosphere, along which opposing air masses contend for possession of the underlying region. And just as military fronts periodically erupt into battle after periods of 'all quiet', so atmospheric fronts, while sometimes quiescent, are more often belts of disturbed weather. As such they make a most important contribution to British weather, being with us, as was shown in Chapter 5, for something like one-eighth of the time.

The concept of fronts has changed somewhat in the 50 or so years since it was formulated by the Norwegians. In part this reflects the increasing importance of upper-air observations during that period and their inevitable impact on ideas based mainly on surface reports. It is now evident that no two fronts are identical but it remains useful to think in terms of a 'model' which all real fronts resemble more or less. The model itself has become more complex and in fact there are several rather than one.

Air Mass Boundaries

The military analogy recedes a little when we examine in detail the relationship between adjacent air masses at a front, but another analogy may be useful. If we contemplate a half-full tumbler of water, we see two fluids, the less dense (air) resting above the denser (water), separated by a horizontal boundary. Colder and warmer air masses are, from this point of view, heavier and lighter fluids respectively and, were the earth a stationary body, the colder air would completely underlie the warmer and the 'front' would again be a horizontal boundary of separation. But the earth is rotating and other forces in addition to gravity are involved. In fact, adjacent air masses are separated by sloping boundaries, the warm partly overlying the cold. We explain this point more fully in Chapter 7 but for the moment may note that we could obtain a somewhat analogous effect by gently swishing the water round in the tumbler, which will give an inclined separating surface lower in the centre than at the sides.

Figure 6.1 attempts to portray the nature of an air-mass boundary in block diagram, plan and cross-section views. The so-called *frontal surface*, seen as a sloping line in cross-section, meets the ground at a line (which to aid confusion is sometimes called the *surface front*) and it is this line that represents the front in plan, i.e. on the synoptic chart. In all diagrams such as Fig. 6.1(a) and (c) we are obliged to exaggerate the frontal slope many times for in reality the angle of inclination is always very small, a degree or two at most. Expressed as gradients, frontal slopes vary from about 1 in 25 to 1 in 300, with 1 in 100 as a useful value to remember in our latitudes.

While we speak conventionally of frontal surfaces, it would be quite wrong to visualize these as sharp planes of discontinuity in the atmos-

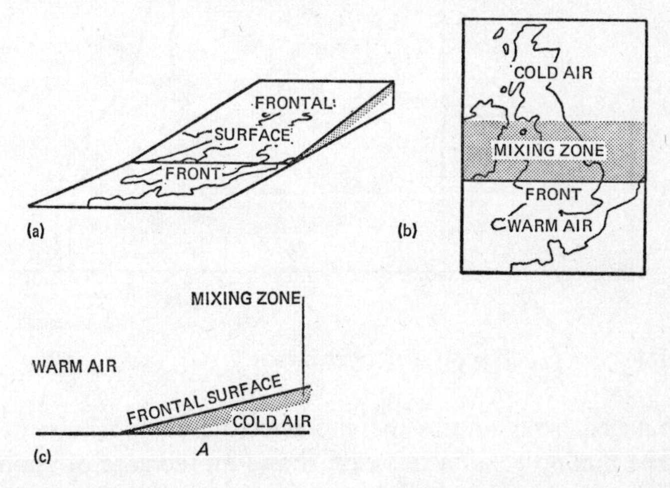

6.1 Idealized structure of a front seen (a) in block diagram (b) in plan (c) in cross-section.

phere. A radio-sonde balloon sent up from point *A* in Fig. 6.1 passes first through the cold air, then through a transitional layer one or two km thick caused by turbulent mixing (the *mixing zone*), and then finally into the warm air above. In Fig. 6.2 an actual ascent through

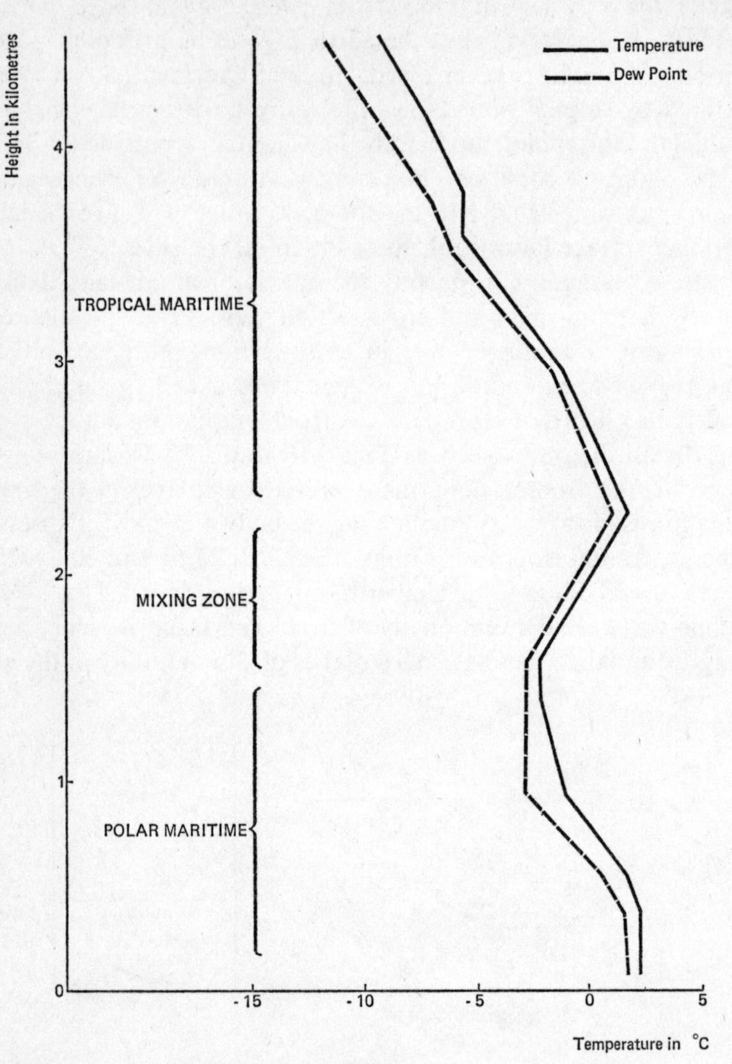

6.2 *Aerological sounding through a front.*

a well-marked front reveals the characteristic composite temperature curve. The mixing zone in this case shows an increase of temperature with height (*frontal inversion*) but in less clear-cut examples is represented merely by an isothermal layer or one with a much reduced lapse rate. The characteristic moistness of the air reflects the cloud and rain of an active front (of which more shortly).

What we loosely call the 'line' of the surface front is therefore really a belt of substantial width, up to 200 km (125 miles) depending on the frontal slope. For this reason we must not expect abrupt changes of properties at the passage of a front, although some are more sharply defined than others. However, on synoptic charts it is convenient to represent fronts as lines rather than belts for practical purposes of analysis and forecasting. It will be seen from Fig. 6.1 that the mixing zone is conventionally regarded as belonging to the cold air.

Active and Inactive Fronts

Air-mass boundaries do not necessarily create bad weather: there are situations in which the two air masses lie passively side by side or flow in parallel streams producing little in the way of frontal weather. For a front to be *active* there must be a movement of either air mass across the front, causing a congestion of air which can be relieved only by removal upward. This is a particular case of *convergence*, a term used by the meteorologist in a precise way, involving both horizontal and vertical motions (see Chapter 7). The simplest frontal model has the warmer (lighter) air rising above the shallow wedge of colder (denser) air but this is not a necessary condition: part of the cold air could ascend also but usually this air would be too dry to give much cloud and so it is effectively the ascent of warm air that gives the front its character and the stability and moisture conditions of the warm air mass that determine the possibilities of frontal cloud and precipitation. Without the essential convergence the front is *inactive* and significant only as a boundary zone.

There are parts of the world where air flow patterns persistently confront air masses of different properties and other areas where such conditions exist temporarily: in other words there are both general and local frontal belts. The term *frontogenesis* denotes processes that generate new fronts or intensify existing ones. Other air flow patterns lead to the decay of fronts (*frontolysis*). Figure 6.3 suggests a situation that would encourage frontogenesis in one region and frontolysis in another: in the former, a convergent flow sharpens the transition zone between the two air masses and leads to frontal ascent, in the latter, a *divergent* pattern leaves a deficit of air that necessitates descent from higher levels and inevitably blurs the frontal contrast. So it is that frontal vigour may vary along the length of a single front from well-defined and strongly active to diffuse and difficult of identification.

As far as the major frontal belts are concerned, the requisite frontogenetic conditions stem from the general circulation (see Chapter 7), which provides regions of low pressure towards which air from different sources is drawn. The nearest such region to affect us is the North

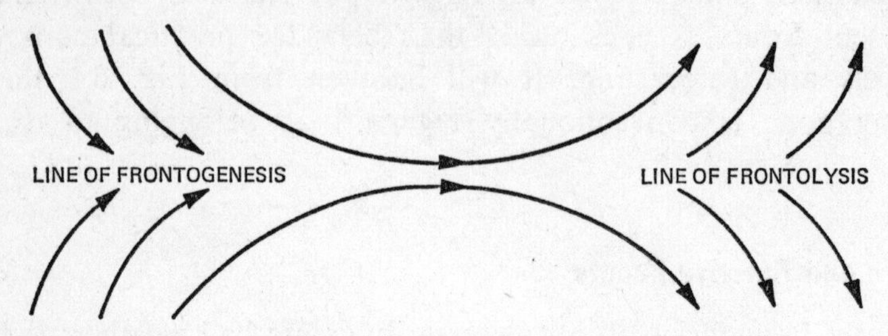

6.3 Air flow patterns favouring frontogenesis and frontolysis

Atlantic in middle latitudes, where a frequent confrontation of Polar and Tropical air masses creates the *Polar Front*. Where such a frontal belt coincides with parts of the earth's surface where sharp temperature gradients occur, frontal contrasts are apt to be intensified (frontogenesis). Such favoured areas include the boundaries of warm and cold ocean currents, for example, in relation to the Polar Front, the Gulf Stream and Labrador Current off the east coast of North America. The fringes of ice-caps or snow-covered land (e.g. the Greenland or Antarctic plateaus) may perform a similar function.

In contrast, the decay of fronts is encouraged by a modification from below that reduces the air-mass contrast (for instance if the two air masses lie stagnant side by side over a uniform surface) or by descending air. This is characteristic of anticyclones and it can be seen from Fig. 7.2 that the breaks in the front (i.e. the frontolyzed portions) occur in the regions of high pressure: the frontal cloud cannot survive long in conditions of sinking (i.e. warming and drying) air.

Warm and Cold Fronts

More often than not, fronts are on the move, in the sense that the entire boundary zone shifts so that one air mass penetrates into territory lately occupied by the other. The direction of this movement distinguishes fronts as warm or cold. At a warm front, the warm air is advancing and the wedge of cold air retreating, so that an observer

12. Thundery weather: mammatus pouches hanging from an overhead cloud and a bank of altocumulus castellanus in the middle distance.

13. Anticyclonic weather, summer: fair weather cumulus

14. Anticyclonic weather, winter: radiation fog.

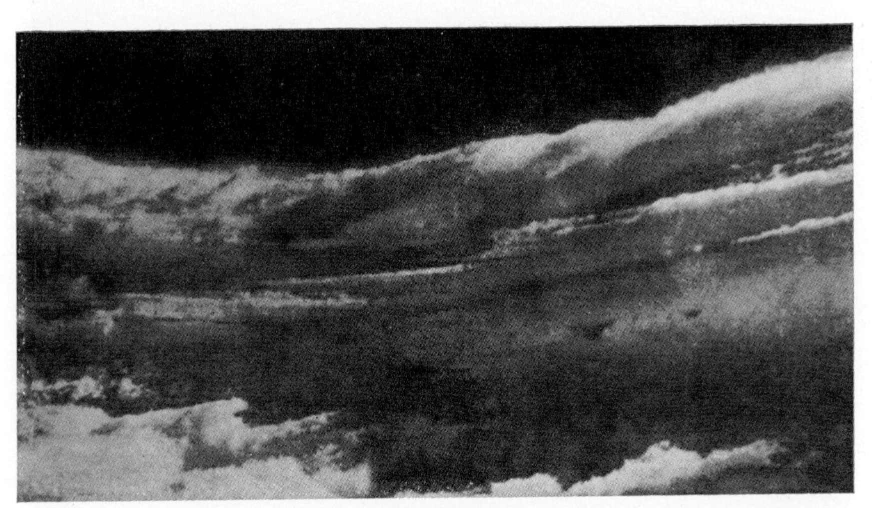

15. Tropical weather: inside the 'eye' of a typhoon. Photograph of the Pacific typhoon 'Marge' taken by Dr. R. H. Simpson from an aircraft at 17,000 feet.

16. A tornado in Britain: the Bedfordshire tornado of 8th June, 1955.

17. Local weather: a solitary hill cumulus above the highest hill. Lleyn Peninsula from near Harlech.

18. Local weather: wave clouds over Dartmoor suggest the shapes of the land-forms producing them.

standing initially in cold air later finds himself in warm air. At a cold front it is the cold air that advances, covering ground yielded by the warm. Fronts with no movement cannot be classed as warm or cold: they are simply stationary or, in practice, nearly so (*quasi-stationary*). A front may be warm in one sector and cold in another (see, for example, Fig. 7.2), which implies a pivotal movement about a quasi-stationary point. It is also possible for a front to reverse its direction of movement and thus change from warm to cold or cold to warm.

A moving front is not necessarily an active one. The required relative motion must be combined with the overall motion. With an active warm front, the warm air encroaches upon the cold and, if the ensuing movement is not quite the gentle gliding of warm air up the frontal slope that was suggested by the earliest model, at least there is considerable ascent of warm air in the vicinity of the front. At an active cold front, we see more readily that the advancing cold air acts somewhat as a wedge undercutting the warm air and forcing it to rise. It might be thought, since basically warm and cold fronts differ only in their direction of movement, that their structures and associated weather systems are identical (apart from being reversed). However the differences in air movement outlined above may be reflected in some marked differences in cloud and precipitation. Before considering some of the further variations in frontal structures that have been identified over the years, we learn what we can from a more detailed examination of two particular frontal situations.

Portrait of a Warm Front

In Fig. 6.4 a warm front showing many usual characteristics is depicted in both weather map and cross-section views. The synoptic chart shows the surface front just reaching the west coast of Ireland. It separates cool Pm air with temperatures as low as 1° or 2°C well to the east of the map from mild Tm, 10°C or warmer, in the extreme west. The greater detail of the fully plotted chart reveals that the cloudy skies over East Anglia are partly due to high cloud associated with the front (though there is also some patchy low cloud belonging to the Pm air) so that we may speak of a frontal system at least 700 km (435 miles) in width. The rain belt has a width of 250 to 300 km (say 150–200 miles). Again, detailed observations show that the rain just affecting North Wales is light and intermittent in character but within 150 km (90 miles) of the surface front is mainly moderate and continuous. Behind it the rain has ceased and the ship report shows cloudy skies and the highest temperature on the chart.

The cross-section, which helps to complete a three-dimensional picture, has been derived from both upper-air and surface data. The frontal slope, determined from the available radio-sonde ascents (see, for example, Fig. 6.2), is about 1 in 200, a very small inclination that is typical of warm fronts. The run of the isotherms shows clearly the

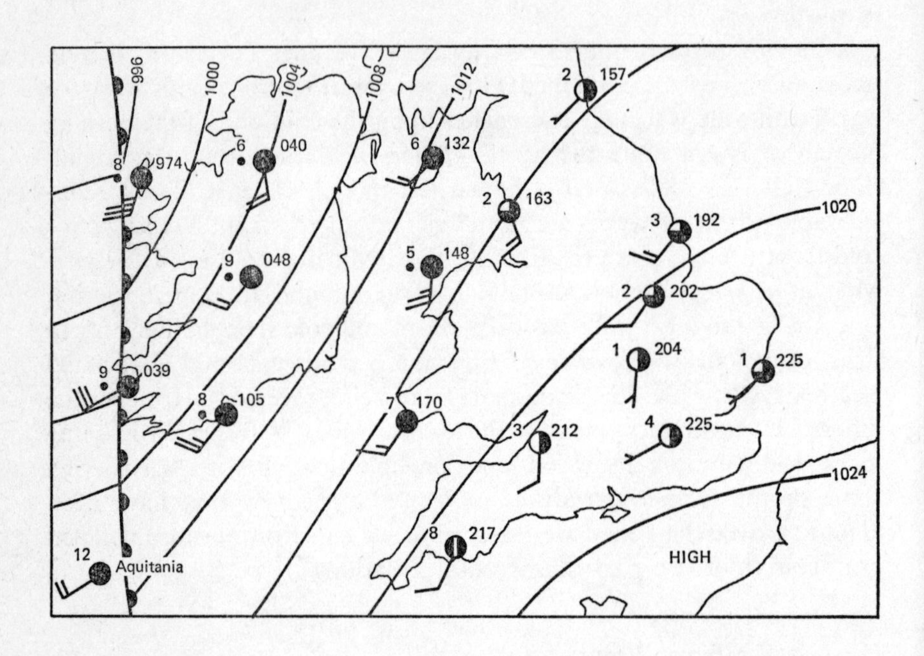

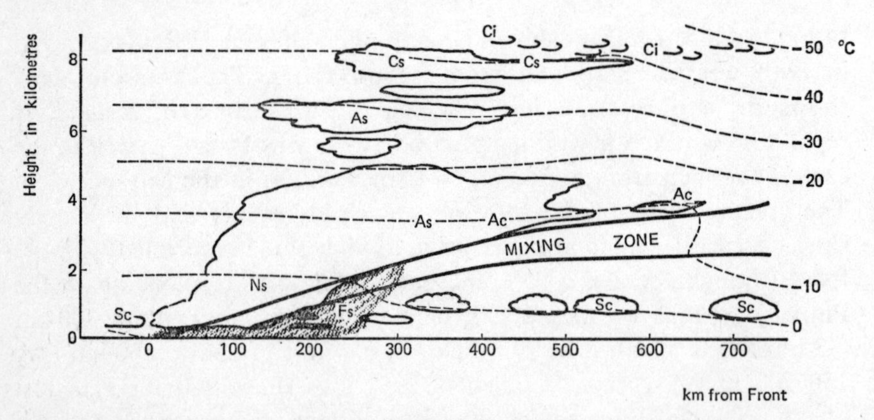

6.4 Portrait of a warm front, 5 April, 1947, 0600 hr GMT. In the cross-section (below), broken lines are isotherms.

temperature change between the two air masses and their downward dip in crossing from warm to cold denotes the mixing zone.

The frontal cloud (as indicated broadly by the humidity data) extends vertically to about 9 km (nearly 30 000 ft) but is solid only to about 5 km (16 000 ft), lying in several thin layers above this. Such a layered frontal cloud, reflecting variations in moisture content or stability conditions at different levels in the warm air mass, is more common than the conventionally portrayed cloud sheet solid to the tropopause (though this may occur with very moist or unstable warm air). It will be seen that the cloud base does not coincide very neatly with the mixing zone. Near the surface front the cloud base has built downwards in the rain-soaked cold air and ahead of this there are ragged patches of very low cloud. With some slow-moving fronts, condensation in the saturated air occurs down to ground level, giving a narrow belt of fog which travels with the front and is known as *frontal fog*. Above about 3 km (10 000 ft) the cloud seems quite divorced from the base of the warm air: recent investigations have shown this to be usual and have revealed the presence of very dry air in and near the mixing zone (which has been interpreted as meaning that air actually subsides into this region from still higher levels).

As such a system moves, a suitably placed observer will experience a characteristic sequence of weather events, including the succession of cloud types which are indicated on the lower diagram of Fig. 6.4. Our observer at, say, Great Yarmouth will already be aware of the top-most clouds of the sequence. At the low temperatures of these upper tropospheric levels, the air contains little moisture and condensation takes the form of thin clouds of ice particles: these are the feathery or thread-like wisps known as *cirrus* (Ci on the diagram: also see Plate 9). Cirrus ('mare's tails') is not necessarily frontal. But if it appears, with time, to consolidate into a denser uniform sheet of ice cloud, *cirrostratus* (Cs), then the warning of an impending warm front is not to be ignored. Cirrostratus is easily recognizable because, being only a filmy sheet, it allows the sun or moon to shine through quite brightly but typically encircled by a *halo*. Solar and lunar haloes are wide luminous rings (subtending an angle of 22° between the sun or moon and the ring) sometimes showing vague colours through dark glasses: they are caused by the *refraction* (bending) of light rays by ice particles. Only thin continuous ice clouds can produce this optical effect. The halo should be distinguished from the *corona*, a smaller coloured ring sometimes seen fringing the sun or moon but always through thin water clouds with small droplets: this is a diffraction phenomenon caused by the mutual interference of the light waves.

The next clouds to appear are of medium elevation (prefix *alto*), usually of the sheet type, altostratus (As) but often, as in Fig. 6.4, mixed with the less stable form altocumulus (Ac), thickening progressively in the direction of the surface front. Altostratus is a quite featureless sheet, allowing the sun to shine through in vague watery fashion while it is thin, but soon thickening and blotting it out entirely. Altocumulus is distinguished by its more definite form. These medium clouds contain both ice particles and supercooled water droplets and so it is from here that the first precipitation will fall. This is at first slight and evaporates in the drier air below: eventually however it reaches the ground as snow or rain. Thick rain-bearing altostratus, with its base lowering to perhaps only a few hundred metres, is called *nimbostratus* (Ns). This murky wet sky, with its *fractostratus* patches scudding along beneath the main cloud base, is too familiar to require description. The rain, usually persistent rather than heavy, giving the ground a steady soaking, eventually peters out: the front is now through, the observer finds himself in the warm air and air-mass weather replaces frontal weather.

In addition to these readily observed changes, self-recording instruments in the path of an advancing warm front register (1) an increase in surface temperature (2) an increase in relative humidity and dew-point (further raised as a result of precipitation) (3) falling pressure, since the front is accompanied by more or less of a trough of low pressure (see p. 149) and (4) an increasing and *backing* wind (i.e. one changing direction in an anti-clockwise sense), which is related to the convergent motion. Following the passage of the front, temperature and humidity become representative of the warm air mass, pressure either steadies or falls more slowly and the wind *veers* (changes in a clockwise direction). Continuous traces illustrating these changes are shown in Fig. 6.10. The art of identifying and locating fronts on the synoptic chart depends on the successful 'reading' of these transitions from the plotted data.

The duration of this or any frontal passage depends on the width of the belt and the speed of movement. As will be shown in Chapter 11 it is possible to gauge the movement of fronts from the isobar spacing: this front is moving east at something like 35 m.p.h. or 16 m/sec, which means that the entire frontal sequence, from early warning cirrus to the cessation of rain, would pass over in about twelve hours and the rain belt in about six. But fronts may move at 50 m.p.h. or very slowly and frontal weather may be short-lived or protracted.

Portrait of a Cold Front

Similarly portrayed in Fig. 6.5 is a cold front, moving southward over the country and replacing mild Tm weather brought by a westerly flow with a cool northerly polar airstream: the general temperature contrast is about 7°C (13°F). The synoptic chart shows that the rain

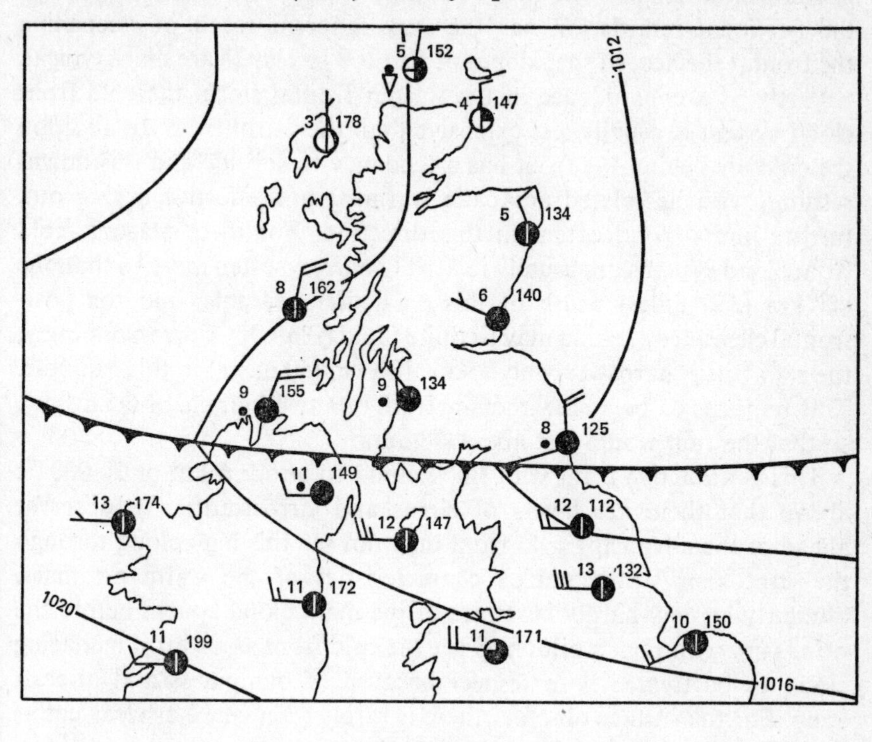

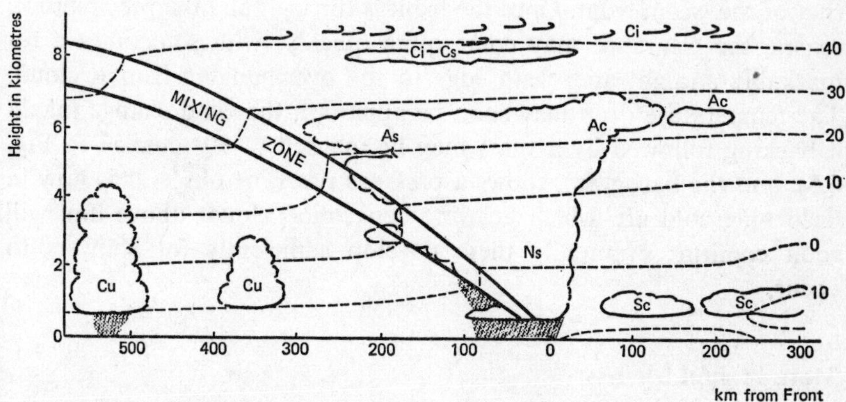

6.5 Portrait of a cold front, 26 October, 1955, 0600 hr GMT. In the cross-section (below), broken lines are isotherms.

belt is mainly post-frontal. The cross-section reveals a much steeper frontal slope (in this example about 1 in 60) than in the corresponding warm front diagram: this is usually the case. It is partly explained by the frictional drag necessarily exerted by the ground on the over-lying air flow. With the warm front this effect further flattens the already gentle frontal slope in its lower portion: in the case of the cold front the frictional retardation has the very different result of steepening the frontal surface, so that slopes of 1 in 100 to 1 in 25 are more typical.

Partly as a consequence of the steeper frontal slope, the cold front cloud system is usually less extensive than the warm front. In addition the cold air behind the front has a tendency to subside and this down-settling, with its related adiabatic warming and effective drying out, further limits cloud extent in this direction. For these reasons, cold front cloud systems are usually relatively narrow, often more so than the 400 km (250 miles) width of this particular example, and the post-frontal clearance of cloud may be quite rapid (Plate X). Correspondingly, the rain belt is narrower, only about 100 km (60 miles) in this example. This happens to be a slow-moving front (about 14 m.p.h. or 6·5 m/sec), so that the rain would last about 4 hours.

This is an active front with thick cloud to about 6 km or 20 000 ft: above that there are layers of cirrus and cirrostratus. An observer ahead of the advancing cold front may not see this high cloud through the stratus or stratocumulus characteristics of the warm air mass. Similarly he may hardly be aware of the main cloud system before the rain is upon him but a pilot may see the cold front as a wall of menacing cloud to be treated with respect because of bumpiness and aircraft icing. For the surface observer there is rarely a sequence as clear-cut as that of the warm front. Once the front is through and the precipitation ceases, the clearance may come quite quickly, with sometimes a remarkably straight and clean edge to the over-hanging frontal cloud. The temperature drop may soon be apparent, the wind change (again a backing followed by a veer) may be marked (a 90° change in Fig. 6.5), and the barograph shows a pressure rise. The observer is now in deepening cold air, which growing convection clouds above him will soon confirm: eventually these develop sufficiently for showers to occur.

More Frontal Models

It was the Norwegian Bergeron who first suggested, in the 1930s, a further division of fronts into what he called *ana-fronts* and *kata-*

fronts. At an ana-front, the relative motion is such that the front is active at all levels so that the warm air undergoes general uplift and the frontal cloud may extend upward to the tropopause. At a kata-front, the active element is confined to the lower few kilometres and only here is the warm air ascending, while above this it is actually sinking and this subsidence is marked by clear air: a subsidence inversion separates the levels of upward and downward movement (Fig. 6.6).

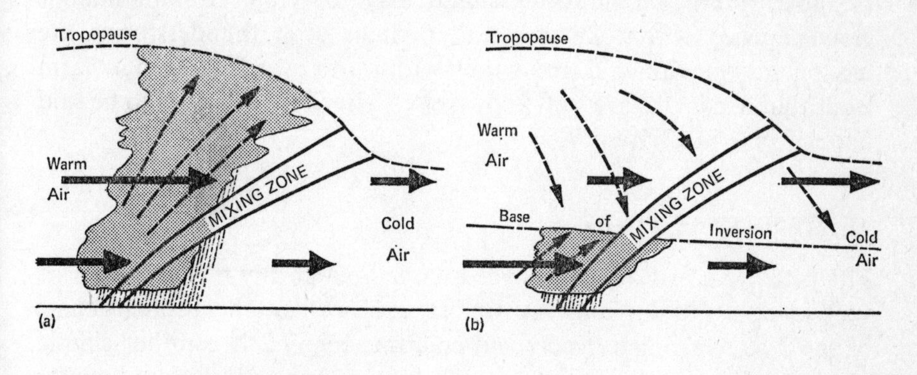

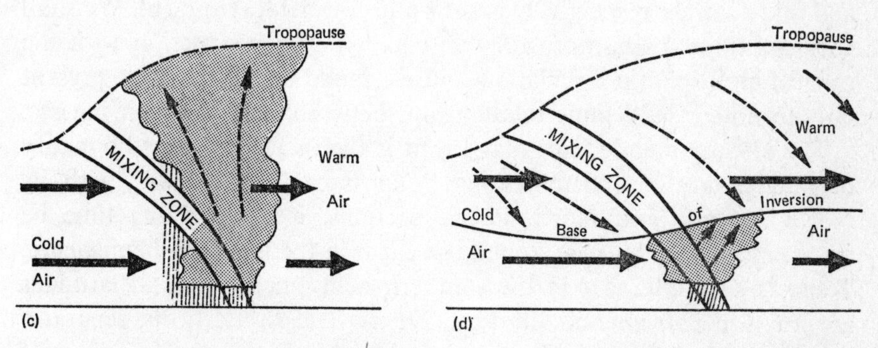

6.6 Ana-fronts (a, c) and Kata-fronts (b, d): thick arrows (length proportional to speed) denote horizontal motions: broken arrows suggest vertical motions. (Mainly after Pedgley.)

Both our portraits were of ana-fronts. The *ana-warm front* is most like the textbook model and is the more common warm type over the British Isles. The *kata-warm front* has a cloud belt of perhaps only 3 or 4 km, virtually only thick stratocumulus, from which a little light rain or drizzle may fall. The *ana-cold front* is the nearest to a mirror-image of the ana-warm, except that the frontal slope is steeper. Rather more common over Britain is the *kata-cold front* with a very restricted cloud belt horizontally and vertically and little precipitation, if any. The main difference between ana- and kata-fronts is one of activity: the ana structure seems to belong to a younger and more vigorous phase of frontal activity, the kata to an older, decaying phase.

These further models do not exhaust the variety of frontal structures. Much depends on the properties of the warm air. Usually over the British Isles, this is fairly stable and the result of frontal uplift is sheet cloud, though this may be thick, but if the warm air is conditionally unstable, more cumuliform cloud may develop. Cumulonimbus clouds, giving heavy showers and perhaps even thunderstorms are more common with cold fronts than with warm over Britain, but warm front thunderstorms are well known elsewhere. There is more to be said about fronts in Chapter 7.

The Major Frontal Belts

The actual pattern of fronts over a large area at any moment may be very complex, as a glance at almost any northern hemisphere chart of the *Daily Weather Report* will confirm: Fig. 7.2 is a rather simple example. The average location of the major fronts during mid-winter and mid-summer respectively is shown in Fig. 7.14(a) and (b). We shall consider these maps more fully in Chapter 7 in connection with the general atmospheric circulation and we make use of them at present only to note briefly some relationships between fronts and air masses.

The already-mentioned Polar Front is the major air-mass boundary of middle latitudes, demarcating Polar from Tropical air: with its North Atlantic and North Pacific sections, it may at any time be traced more than halfway round the earth in the northern hemisphere. There is a counterpart in the southern hemisphere. In high latitudes are found the *Arctic* and *Antarctic Fronts*. The Arctic Front separates Arctic (direct Pm) air from less fresh Pm, that is to say, air masses of similar origin distinguished by different degrees of modification: the air-mass contrast is not usually as strong as in the case of the Polar Front, so that the Arctic Front may be less well marked a feature. The Antarctic Front however divides very cold Pc air originating over the Antarctic ice-cap from much warmer Pm further to the north.

In low latitudes we find a feature described on the maps as the *Intertropical Convergence Zone* (ITCZ), although it has also been called at various times the Intertropical Front (ITF) and the Equatorial Front. If the essence of a front is an air-mass contrast, then this is clearly lacking over large stretches of equatorial ocean where very similar air (historically Tm) arrives from both north and south. Only over or near land, where Tc air may confront Tm, can the requisite contrast develop and only well away from the Equator (as in monsoon circulations) can a frontal structure resembling that of middle latitudes form (see Chapter 8). The term ITCZ begs least questions and there is certainly a more or less broad zone of convergence in these latitudes, extensive though not continuous, with important meteorological consequences.

The mid-winter and mid-summer pictures differ mainly in the position of the fronts which tend to shift latitudinally with the sun. They also generally weaken in the summer season. Otherwise, the only striking difference between the two maps is the existence of what is in effect a section of the Polar Front in the Mediterranean region in winter. This *Mediterranean Front* forms between cold Polar air (often Pc from the Eurasian interior, sometimes Pm or A from further west) and warm air, some of it Tc from North Africa, but some of it much-modified 'Mediterranean' air of varied origin, warmed by long residence over the sea. No such clash of air masses can occur here in summer.

Frontal Disturbances

From the time that weather maps were first drawn, it was recognized that there were regions of low pressure – *depressions* or *cyclones* – which could be enclosed by roughly concentric isobars and seemed to be coincident with regions of bad weather. An early 'cyclone' model (of 1888) shows basically such an isobar pattern over-written with descriptive terms such as 'mare's tails', 'halo', 'watery sun', 'dirty sky', 'line of squalls', which should have a familiar ring to the reader who has persisted this far in the present chapter. It was part of the Norwegian achievement, during the First World War, to recognize (not without perceptive hints from earlier workers) that what seemed a rather random distribution of weather phenomena could be well explained by superimposing a changing pattern of air masses and fronts on to the old picture of the depression. In the Bjerknes scheme of things, warm and cold fronts fall into place as essential components of a frontal depression or *disturbance*.

Figure 6.7 shows such a system. The concentric isobars are associated with a characteristic *cyclonic* wind flow, i.e. anticlockwise in the northern hemisphere (more of which in Chapter 7). The warm and cold fronts are linked at the centre of the low pressure pattern and demarcate a roughly triangular area plainly of the warmer air mass (the *warm sector*): on all other sides, there is cold air. The distribution of weather is what would be expected in such a composite system which embraces two air masses and two fronts, with the addition that the frontal belts of cloud and rain merge near the centre into a generally cloudy and rainy area.

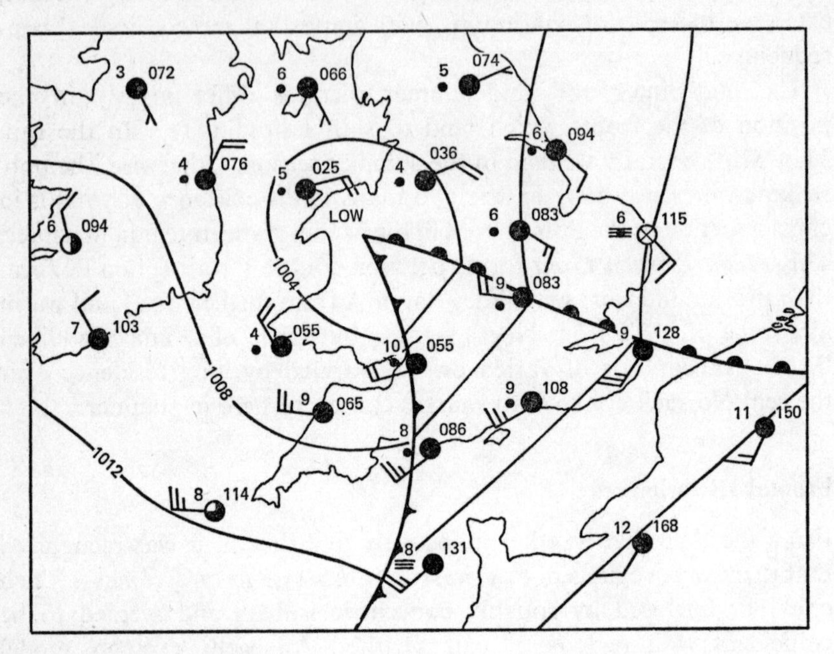

6.7 *A frontal depression.*

We have only to glance at a few *Daily Weather Report* maps to be convinced that frontal disturbances are commonplace features of our own latitudes. Closer examination will however show that Fig. 6.7 depicts only one stage in a sequence of development, which we now illustrate with an actual example extracted from DWR charts – the life history of 'Low J' (in British synoptic practice, weather systems are named alphabetically). This is the process known as *cyclogenesis.*

The History of 'Low J'

In Fig. 6.8(a) we see part of the Polar Front out in the western Atlantic, separating Pm air on the north-western side from Tm to the south-east. It is straight and undisturbed: the fact that it is shown as a cold front means that there is a southward movement but the indeterminate

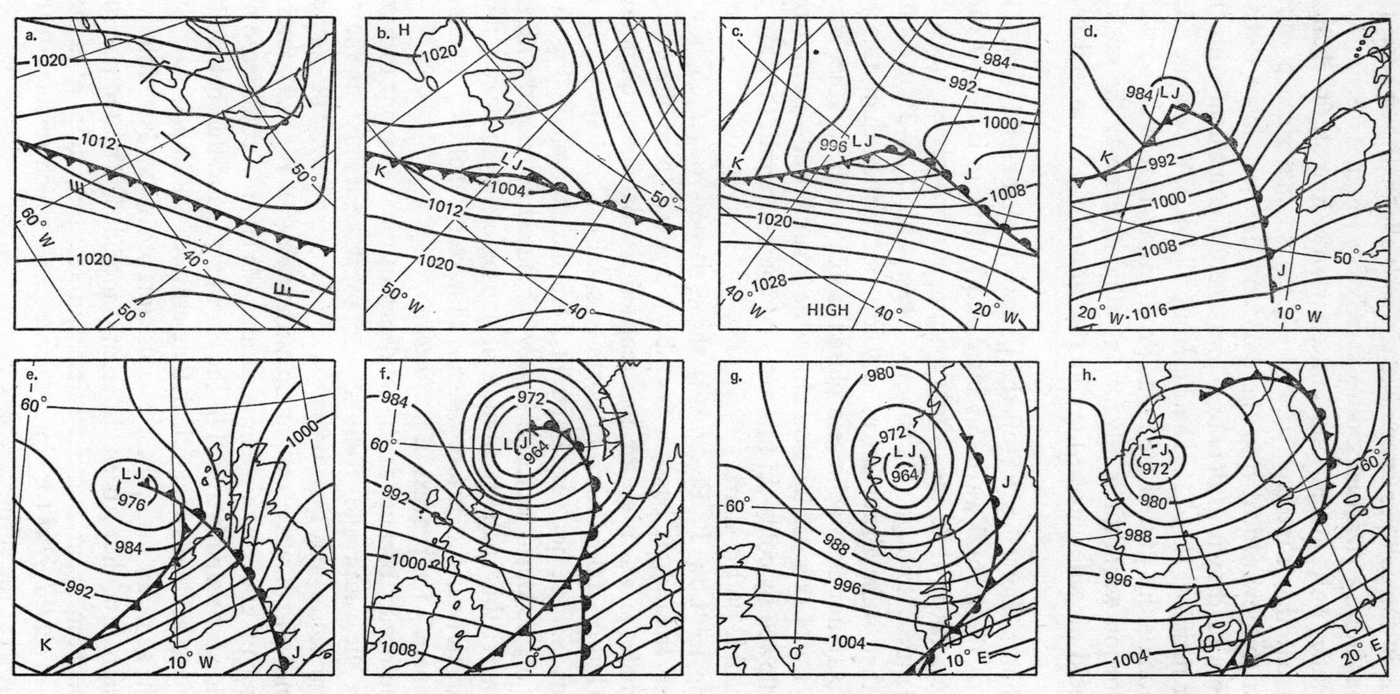

6.8 *The history of Low 'J' (a) 9 November, 1954 1200 hr (b) 10 November, 0000 hr (c) 10 November, 1200 hr (d) 11 November, 0000 hr (e) 11 November, 0600 hr (f) 11 November, 1800 hr (g) 12 November, 0000 hr (h) 12 November, 0600 hr (all times GMT).*

wind arrows in the cold air suggest that this is slight. In contrast, the warm air has quite a vigorous south-westerly flow. Figure 6.8(b) shows the same scene twelve hours later and clearly two important changes have occurred: the front has 'waved' slightly so that surface warm air forms a shallow salient into the cold and the two closed isobars that can now be drawn in denote a pressure fall coincident with the wave – our incipient 'Low J'. The superimposition of a cyclonic eddy or circulation on this part of the Polar Front means that in the forward (eastern) portion warm air is advancing, while in the rear the push is from the cold air side: hence the differentiation into warm and cold fronts.

Twelve hours later (Fig. 6.8(c)) the pattern is well established. The amplitude of the frontal wave has increased and there is now a wide warm sector. Meanwhile the whole system has progressed eastward to mid-Atlantic. Although this is not the case with 'Low J', it should be noted that some disturbances develop no further than this (in fact some may not even achieve a closed isobar on the weather map) but simply travel as a corrugation along the Polar Front – warm front weather followed by cold – in this arrested condition until they are no longer traceable on the chart.

Returning to 'Low J', Fig. 6.8(d) shows it now (after 36 hours) approaching Ireland. It has considerably *deepened* (note the central pressure) and the warm sector has reached its maximum amplitude (over 1100 km or 700 miles from north to south). This stage could be described as *mature*, not very different from the example in Fig. 6.7. The width of the warm sector is however reduced compared with the previous chart: this is the beginning of a progressive squeezing aloft of the warm air from the surface which is characteristic of the later stage of development – the process of *occlusion*.

Occlusion occurs because of the tendency of the cold front to move faster than the warm and eventually to overtake it, first at the tip of the warm sector and then progressively further away. The cold air mass has now, as it were, caught up with itself (see Fig. 6.9), forcing the warm air up above a thickening continuous layer of cold dense air. But the cold air originally ahead of the warm front and that behind the cold front are seldom of identical temperatures. Although part and parcel of the same air mass, these two streams of cold air have followed different paths and have been differently modified. The temperature contrast between them manifests itself in the formation of a new front – an *occluded front* (or simply *occlusion*) – which replaces warm and cold front at the surface.

Figure 6.9 shows that there are two possibilities. The cold air sweeping into the rear of the depression from a northerly or north-westerly direction is fresher and colder than that ahead of the warm front and consequently undercuts it just as it had previously undercut the warm air: the resulting surface front is then known as a *cold occlusion* (Fig. 6.9(c)). It could however happen that the air behind the cold front is returning from a south-westerly quarter (Pmr) and is then the warmer of the two cold air streams: on occlusion, a warm front relationship is established (Fig. 6.9(d)), giving a *warm occlusion*.

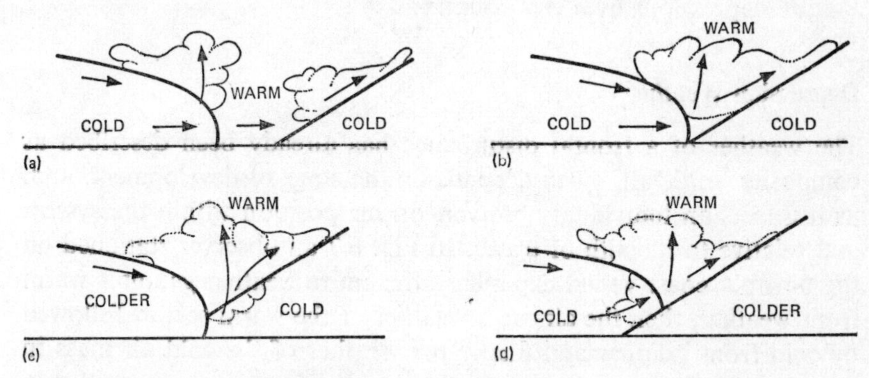

6.9 The process of occlusion (a) well before occlusion (b) just before occlusion (c) cold occlusion (d) warm occlusion.

Returning now to the history of 'Low J', we see the occlusion process at work in Figs. 6.8(c) to (f), in the progressive replacement of two fronts by one and the shrinking of the warm sector. Meanwhile the depression continues to deepen and reaches its lowest central pressure (964 mb) over Norway in Fig. 6.8(g), when occlusion is almost complete. We may appreciate that there is something inherently unstable in a situation that juxtaposes warm light air and cold heavy air at the ground and that effort will be expended to produce a more 'natural' atmosphere stratification. The potential energy implicit in the warm sector stage of the depression becomes available, once occlusion has occurred, to increase the strength of circulation (i.e. is converted to kinetic energy). From this peak of intensity, the system begins to age. The central pressure rises and fewer isobars can be drawn on successive charts. 'Low J' is already beginning to 'fill' in Fig. 6.8(h). In these later stages the occluded front, tightly coiled near the centre of the disturbance, becomes unrecognizable and the air here is virtually homogeneous. Ultimately, all traces of 'Low J' will have vanished from the chart.

The sequence of events just described has taken rather less than three days during which time 'Low J' has moved something like 4800 km (3000 miles), that is at an average speed of about 45 m.p.h. (or 20 m/sec). From the initial waving on the Polar Front to the beginning of occlusion has taken just over a day. This is a fairly typical time schedule. It means that a disturbance that begins life in the western Atlantic, like 'Low J', is likely to be more or less occluded by the time it reaches north-west Europe, though this is not to suggest that we never see mature depressions over this country.

Depression Weather

The weather of a frontal disturbance has already been described as composite: in detail, much depends on the stage of development, and, as it affects an individual observer, on his position within the system and relative to its path of travel. In Fig. 6.7 an observer stationed on the Norfolk coast would experience the entire weather gamut – warm front weather, then the air-mass weather of the warm sector, followed by cold front weather and finally the weather of the cold air mass in the rear. An observer in Newcastle would suffer only a general rain area and cold air all the time. At the other extreme, the fronts weaken well south of the centre and an observer in Belgium might report only frontal cloud and an air-mass change, with little or no rain.

The passage of a frontal system similar to that in Fig. 6.7 is strikingly shown by continuous traces in Fig. 6.10. The rapid air-mass changes completely upset the normal diurnal temperature variation, notably in the sharp temperature drop at about 10 a.m. The relative humidity trace reflects the incidence of rain as much as the air-mass contrast. As sometimes occurs with winter depressions – and very aptly illustrative of the temperature changes involved – the warm front precipitation began as sleet which later turned to rain and the cold front first yielded rain which soon turned to snow.

In an occluded low, there is one surface front rather than two, but the weather pattern may be an amalgam of warm front and cold front sequences, though in somewhat telescoped form. Rain may continue to fall from the cloud in the lifted warm air (from what is sometimes called the *upper front*) in addition to the frontal interactions nearer the surface. For these reasons occlusion weather may be the worst we experience, especially if it is slow-moving. After a few days, however,

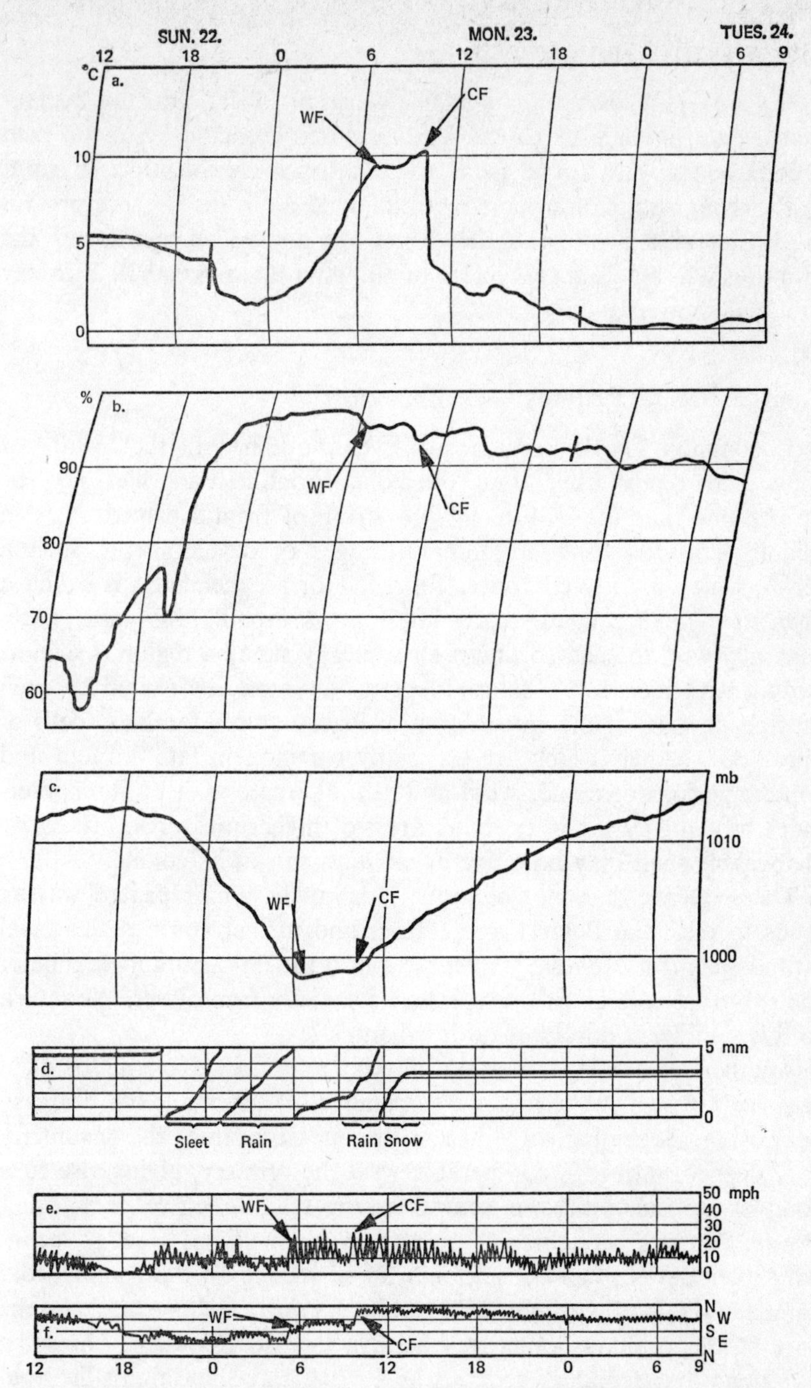

6.10 Weather of a frontal depression, illustrated by continuous traces of (a) temperature (b) relative humidity (c) pressure (d) rainfall (e) wind force and (f) direction, at Abingdon, Berks, 22–24 January, 1956. The arrows marked WF and CF indicate the passage of the warm and cold front respectively.

continued uplift has so denuded the warm air of its moisture content that it contributes little to the weather pattern: a cold occlusion then becomes effectively a cold front and a warm occlusion a warm front. In the dying stages of a depression, the cold air is everywhere present in considerable depth and the fronts are no longer significant: the resulting weather is then typical of the cold air mass, that is to say cold and showery.

Complex Frontal Patterns

It is common for the sequence of events outlined in the history of 'Low J' to repeat itself along the Polar Front so that there may be at any one time (as in Fig. 6.11) a string of frontal disturbances in middle latitudes, often in different stages of development. Such a series, which usually comprises four, five or six members, is called a *depression family*. Each depression follows broadly the same path, generally west to east, so that a strategically situated region – such as Britain with regard to Atlantic depressions – may be visited by each member in turn of the entire family. Hence our too-familiar spells of unsettled weather, which are especially common in late autumn and winter: periods of cloud, wind and rain alternate with brighter interludes brought by the intervening areas of higher pressures. The entire changeable spell may take five or six days to run its course.

The sequence does not continue indefinitely. The repeated waving tends to push the Polar Front further and further south of its usual latitudes until it eventually trespasses into the unfavourable regime of the sub-tropical high pressure, where it quietly frontolyses. We return to this significant development in Chapter 7.

Another common occurrence is the appearance of a so-called *secondary depression* within the circulation of the main (or *primary*) depression. Beginning as a minor low pressure area, the secondary may deepen until it is comparable with the primary, giving rise to a complex system with two centres which tend to circle each other slowly. Patterns with three centres are also sometimes seen. A secondary often develops as a result of renewed waving on the trailing cold front at the rear of a primary (Fig. 6.12(a)): reluctance to occlude on the part of the primary sometimes betrays this development. On some occasions the secondary remains as a wave travelling along the cold front, on others it grows into a fully-fledged depression and this may mark the beginning of a depression family.

Also a favoured area for the formation of a secondary is the tip of

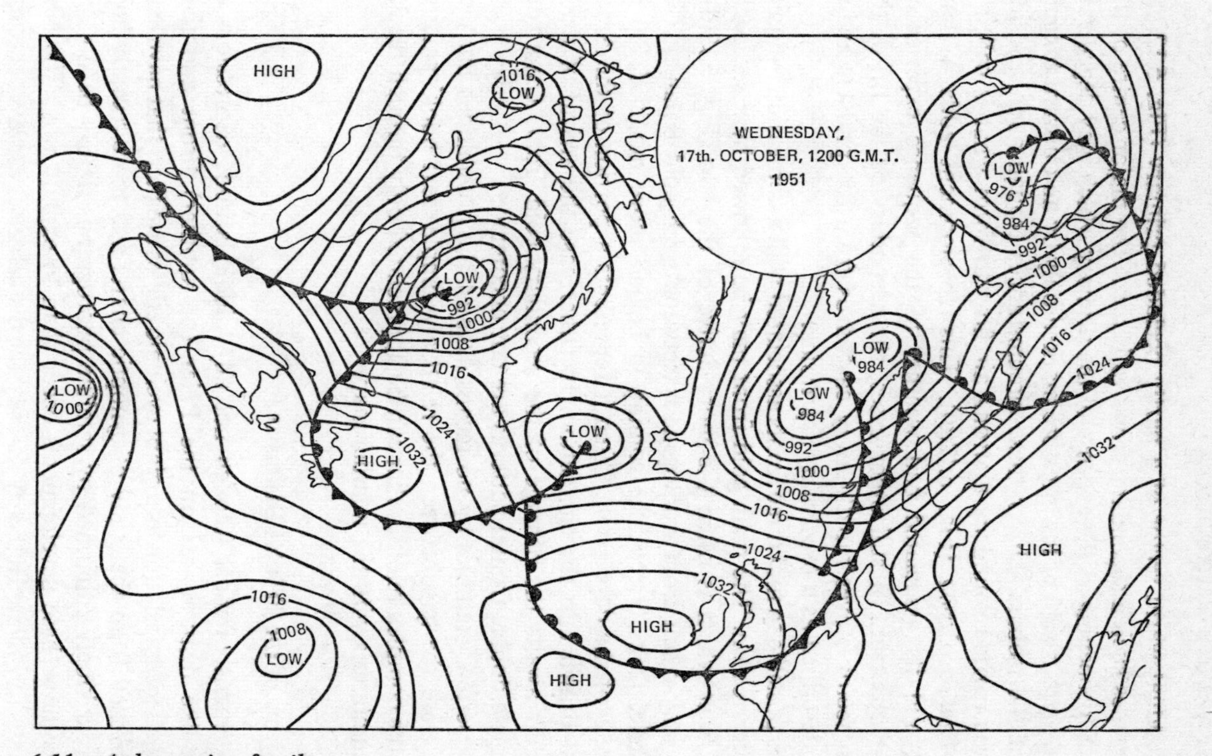

6.11 *A depression family.*

the residual warm sector of a partly occluded primary (Fig. 6.12(b)). Sometimes such a secondary grows to become the major centre of the system, deepening at the expense of the primary which may degenerate into a mere *trough* of low pressure marked only by an outward tongueing of the isobars. Under these circumstances trough and occlusion may be carried round by a strong air flow to form what is called a *back-bent occlusion* (Fig. 6.12(c)). This development adds a third front and its associated bad weather to the sequence experienced by an observer. In still other cases, the secondary at the tip of the warm sector moves rapidly away eastwards from the primary and is then described as a *breakaway depression.*

It is possible to find three fronts associated with a depression for quite another reason. The main cold front may be followed by a *secondary cold front* which is often the more active of the two (Fig. 6.12(d)). This may result from frontogenesis between fresh cold Polar air (for example, from Greenland) coming in behind more modified Pm in the rear of the depression. Often it represents a portion of the Arctic Front drawn into the circulation. In either case the arrival of really cold air awaits the passage of the secondary cold front. Lows with complete double frontal patterns are not uncommon.

These few examples by no means exhaust the possibilities but may help close the gap between the relatively simple 'text book' patterns and the complex realities that can be appreciated only through regular acquaintances with the weather map.

Problems of the Frontal Disturbance

The genesis of most (though not all) middle-latitude depressions as wave-like disturbances on the Polar (or another) Front and their subsequent translation into giant cyclonic eddies are facts that have been established from the study of synoptic charts and more recently have been strikingly confirmed by satellite photographs. It has been easier to see how these developments occur than to explain why. The Norwegians formulated a theory that saw the disturbances as waves corrugating the frontal surface much as wind waves corrugate the surface of the sea. This latter phenomenon depends on a marked change of velocity as between water and air and such a *wind shear* seems to be a necessary requirement for the initiation of disturbances at a front. This has to be a *cyclonic* wind shear with the warm air having a stronger westerly (i.e. eastward) motion than the cold air. Figure 6.8(a) fulfils this requirement since the warm air is moving eastwards while the cold

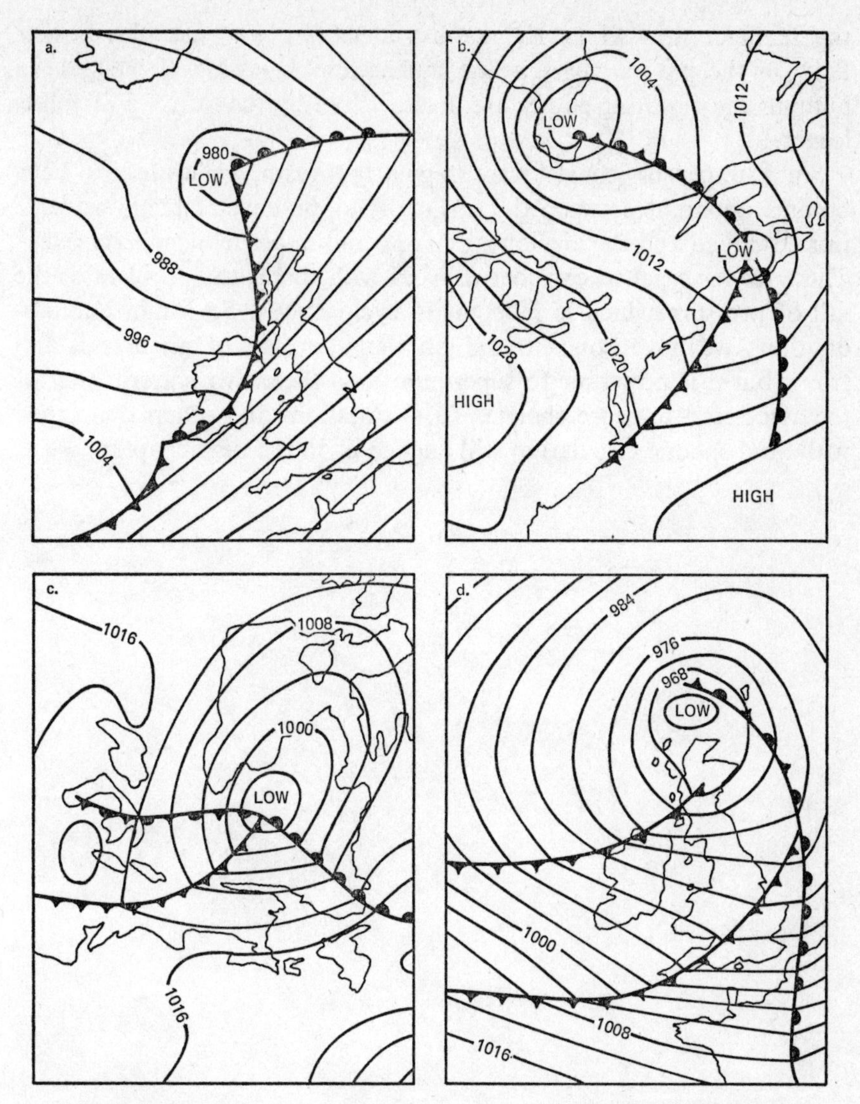

6.12 Complex frontal patterns (a) wave on the cold front of a depression (b) secondary depression at the tip of the warm sector of a partially occluded primary (c) back-bent occlusion (d) secondary cold front. Chapter 7.

air is moving westwards, but a cyclonic wind shear would equally exist if the cold air were also moving eastwards but at a slower speed than the warm air. In these conditions, the eddies readily form in the zone of wind shear. An analogous arrangement springs to mind, when a wind blows parallel to a city street. There is a wind shear between the stronger flow along the unobstructed roadway and the weaker flow on the pavements retarded by the flanking walls or hedges. In autumn the resulting eddies are made visible by the swirling of fallen leaves.

Since upper-air charts were regularly drawn, relationships have become apparent between the surface synoptic developments we have just described and the air flow at middle and upper tropospheric levels. These we have yet to consider. Linked with this is the problem of the fall of pressure which is inherent in cyclogenesis. So far in our descriptions we have concentrated on the interplay of air masses and fronts but it is now time to superimpose on these two patterns that of the associated pressure change. These questions and others concerned with atmospheric circulation will engage us in the next chapter.

7 Atmospheric Circulation

Circulation is a fundamental feature of our atmosphere. On a small scale we see it in the eddies that swirl dust and leaves about on the ground: on a vast scale it dictates the travel of air masses and fronts and thus lies behind the day-to-day fluctuations of our weather. Atmospheric circulation is always a response to inequalities of pressure and the distribution of pressure as shown by the isobar pattern forms the framework of any weather map. Pressure patterns and horizontal air flow (wind) are intimately related, as we shall see: so are horizontal and vertical circulations.

Pressure and Winds – Forces in Balance

Strictly, pressure represents the force exerted by a gas on unit area of an adjacent surface but for meteorological purposes it can be regarded as the weight of overlying air. At sea-level this amounts on average to nearly 15 lb pressing down on every square inch of us (equivalent to a little over 1000 g/cm^2) but this burden is equalized by the pressure of air within our bodies. Except for occasional temporary deafness when pressure changes relatively quickly (as when descending a mountain by funicular railway) we are oblivious of this weather element. Sea-level pressures may rise or fall by 50 mb in a day and, directly, we are unaffected but the significance of such barometric changes for the weather to come is quite another matter.

The wind, on the other hand, is a common part of our everyday weather experience. It influences our personal comfort out of doors and sometimes (in the guise of draughts) indoors, it dries the washing, sails yachts and is a perennial source of power or, in excess, of disaster (see Chapter 12). The relation between pressure and wind is our first concern in this chapter.

By analogy with the gradient of sloping ground, we use the term *pressure gradient* to indicate the change of pressure with distance (strictly, at right angles to the isobars). Gradients may be 'steep' or 'tight', with closely packed isobars, or 'slack', with widely spaced isobars. Wherever there is a pressure gradient we should expect air to flow from high pressure (more air) to low pressure (less air) until the inequality disappears and that the flow should be stronger the steeper the gradient. But if we look at the weather map examples in this book we find that while wind speed indeed depends on the closeness of the isobars, the direction of the wind arrows is not down the pressure gradient, i.e. directly across the isobars, but much more nearly along them. Winds may blow steadily for long periods and we may wonder what maintains the pressure gradient. Without pressure differences there would be no winds but clearly other influences are also involved.

We may approach this by considering the pressure gradient in the vertical direction, which is of course many times greater than the steepest ever encountered in the horizontal plane. However, surface air is prevented from escaping upwards at high velocity by the presence of the countering force of gravity. In other words, two opposing forces are in approximate balance, from which small departures from time to time result in vertical air movements, upwards or downwards.

It is less obvious but nonetheless true that there is a balancing force to be thrown into the reckoning with the horizontal pressure gradient. This force is an inevitable consequence of our spherical spinning Earth. If a piece of tracing paper on which a straight line has been drawn is held against a globe so that the line coincides with say the Greenwich meridian in the northern hemisphere and the globe is slowly rotated west to east while the paper is held pointing in a constant direction, it will be seen that the meridian bends away to the left of the line, or, from another viewpoint, the line increasingly departs to the right of the meridian. We are reminded of the Foucault pendulum which seems to change its direction of swing throughout the day whereas in fact its direction in space is quite constant, while the earth with its reference grid of lines of latitude and longitude rotates away beneath it. Thus it is that a particle of any kind set in horizontal motion on the earth's surface will appear from our earth-bound view-point to turn away to the right of its initial direction of movement if we are in the northern hemisphere (but, as a little thought or further juggling with the globe will show, to the left in the southern).

This apparent force with very real consequences is sometimes called the *Coriolis Force* (after the physicist who defined it quantitatively) or the *deflecting force* due to earth rotation. It is best appreciated through certain analogies, although no one covers all aspects of the situation. A useful example is the gramophone turntable with the spindle representing the north pole and the outer edge the equator. Place a sheet of paper on the turntable, a hole being pierced for the spindle, lay a ruler with one end resting on the spindle and the other on some convenient fixed base outside the turntable, rotate the latter manually anticlockwise and draw a line from the centre outwards along the edge of the ruler: this palpably straight line will appear on the paper as one curving away to the right. (Note that if the turntable is actually switched on, the deflection will be to the left, since turntables rotate clockwise, thus giving in effect a southern hemisphere view of the situation.)

Any point on the earth's surface executes a complete revolution about the axis in 24 hours but the linear distances covered obviously depend on the latitude, a point on the equator travelling at about 1050 m.p.h. (470 m/sec) and one at 60°N at about 500 m.p.h. (about 224 m/sec), while the pole, like the gramophone spindle, simply spins on the spot. The magnitude of the Coriolis deflection, apart from its dependence on the rate at which the earth turns around its axis, is greater the higher the latitude and the stronger the wind speed. The direction of the force is at right angles to that of the actual air motion and to the right in the northern hemisphere (to the left in the southern).

When air pursues a steady straight path, it is assumed that the pressure gradient (PG) and deflection force (DF) are in exact balance, with the alignment shown in the central portion of Fig. 7.1. The push of the pressure gradient is always from high to low pressure, the deflecting force acts in precisely the opposite direction and the resulting wind blows along the isobars with the low pressure to the left. This is the relationship first expressed by Buys Ballot in 1857 in the famous Law well known to every Boy Scout (if we stand with our back to the wind the low pressure is to our left in the northern hemisphere, but to our right in the southern). The wind adjusted to this balance of forces is known as the *geostrophic wind*: its direction is that of the isobars and its speed is proportional to their spacing.

Conditions in the atmosphere are not always geostrophic: indeed they cannot be, or new pressure developments would not be possible. If the pressure pattern changes, the balance is upset but the Coriolis force always strives to readjust to the new pressure gradient and so eventually to restore the balance.

When the wind follows a curved path it is subjected to another compulsion: it does not matter whether we regard this as a force directed outwards from the centre of curvature (*centrifugal force*) or as the equal and opposite inward (*centripetal*) force retaining the air in the system. In the case of *anticyclonic* curvature (flow round a high-pressure system) as can be seen from the left-hand portion of Fig. 7.1, it is the

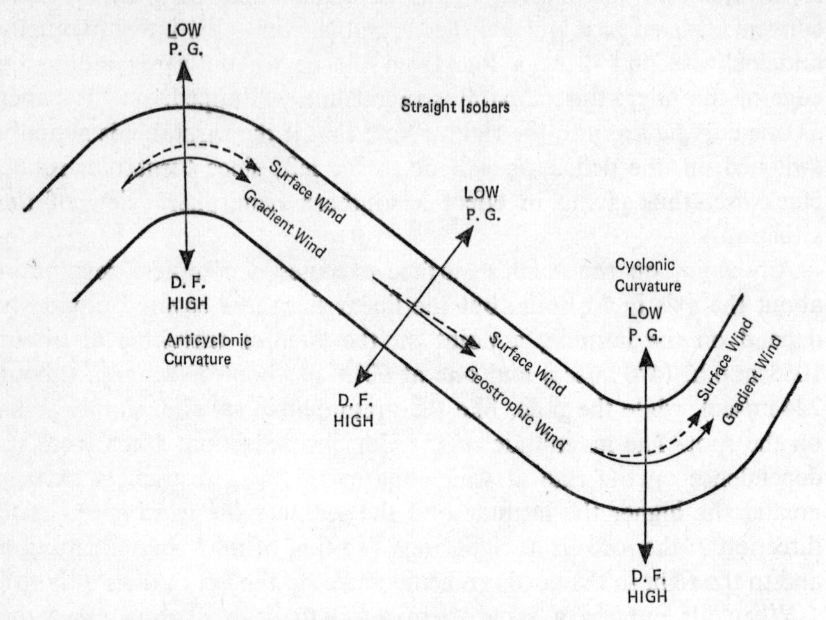

7.1 *Balancing forces involved in horizontal air flow.*

additional task of the deflection force to maintain the air in its curved motion: the deflecting force is thus stronger than would be necessary in a purely geostrophic relationship and the wind speed exceeds the geostrophic value, i.e. is *supergeostrophic*. In the case of *cyclonic* curvature or flow around a low-pressure system (right-hand portion of Fig. 7.1), the pressure gradient force must be the stronger and the wind has less than the geostrophic value, i.e. is *subgeostrophic*. The winds so maintained are called *gradient winds,* rather than geostrophic, but equally they blow along the isobars and are subject to Buys Ballot's Law. We see from the diagram that air circulates clockwise around an anticyclone and anticlockwise around a cyclone and if we forget which is which, the Law can always be invoked to give us infallibly the right answer. (The directions must be reversed for the southern hemisphere.) The magnitude of the centrifugal/centripetal factor is small except with very pronounced curvature of the flow.

It is important to remember that the geostrophic balance is always disturbed near the ground, where friction against a more or less rough surface has the effect of slowing down the moving air. This weakens the deflecting force, which is partly dependent on the wind velocity, and the pressure gradient exerts its superiority by causing the air to flow somewhat towards the lower pressure (Fig. 7.1). All but very light surface winds in fact blow at a small angle to the isobars from high to low pressure, as reference to any weather map will show: the frictional force joins the other two to produce a new balance.

In general the effect of surface friction is felt in the lowest 500 m (1640 ft) or so of atmosphere – the so-called *friction layer*. At about 500–1000 m the wind is in fact adjusted to the isobars (i.e. we can speak of the geostrophic or gradient wind), assuming that the pressure field is unchanging. The surface wind (usually measured at 10 m) differs from the undisturbed wind both in speed and direction, to an extent depending on the roughness of the surface. This is generally least over the sea where the surface wind may have two-thirds the speed of the 500 m wind and be backed from it by only 10°. Inland, during the day, the surface wind is nearer half the 500 m wind and backed by 20° to 30°. At night the surface wind generally drops as turbulence dies down, no longer transporting the faster velocities of higher levels down towards the ground, the speed reducing to perhaps only a fifth of geostrophic and the angle increasing to 40° or 50°.

Common Pressure Patterns

With these principles in mind, we may identify the common pressure patterns which are all illustrated in Fig. 7.2, a fairly typical northern hemisphere chart. We find extensive areas of straight isobars. Elsewhere are regions of low pressure, some with well-defined centres enclosed by isobars (depressions, cyclones, 'lows'), some mere extensions or *troughs* of lower pressure: sometimes these are associated with fronts (*frontal troughs*), formerly called 'V-shaped depressions' because of the characteristic isobar pattern. The high pressure centres (anticyclones, 'highs') bulge out in *ridges* or *wedges*. The least determinate pattern is the *col*: this is a kind of no man's land between two highs and two lows, a region of stagnant air, for which reason it is often a favoured area for fogs in winter and thunderstorms in summer.

Upper Winds

For reasons earlier discussed, the surface wind is commonly backed

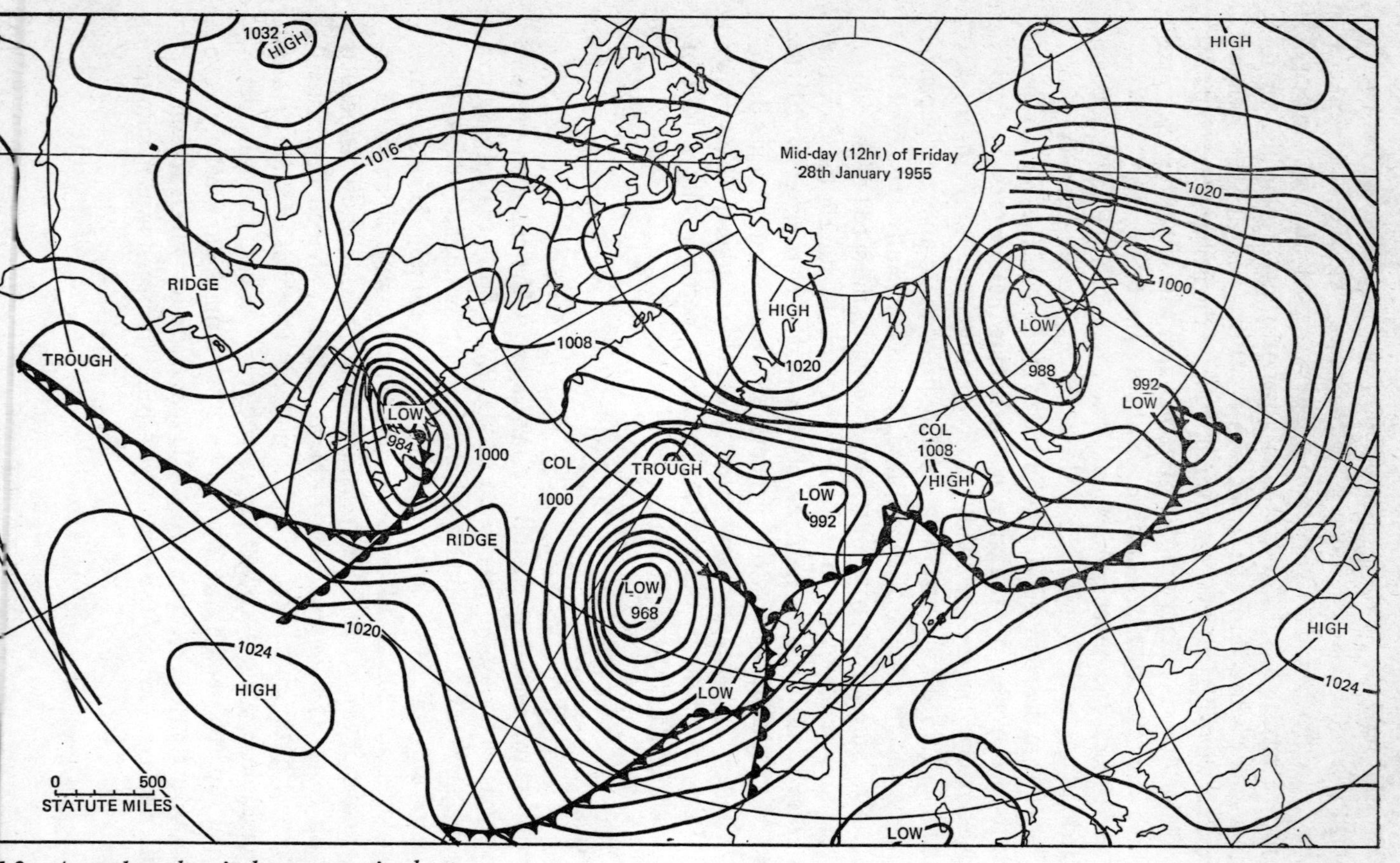

7.2 *A northern hemisphere synoptic chart.*

from the 500 m (geostrophic or gradient) wind by up to 30° inland and during the day, a difference often clearly seen in the comparison of the direction of a wind vane or a smoke plume with that of low convection cloud. At higher tropospheric levels, the drift of cirrus cloud may show yet another direction. We may begin to see why with the help of Fig. 7.3.

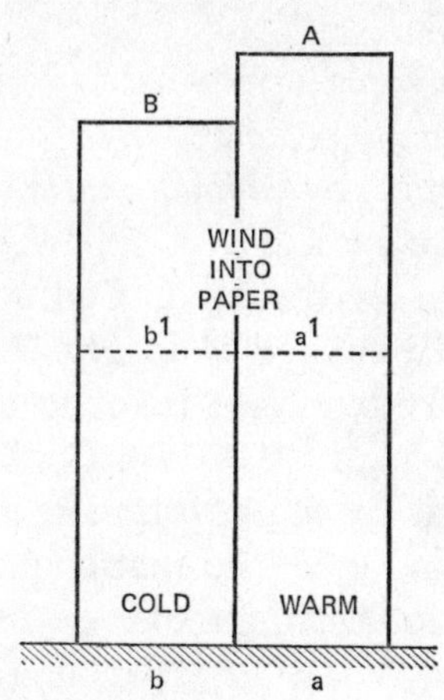

7.3 Illustrating the principle of thermal winds.

This diagram shows two adjacent air columns *A* and *B*, each containing the same weight of air (so that the surface pressure at a equals that at b), but column *A* is warmer throughout than column *B*. The air in column *A* is consequently less dense than in column *B* (which must then be the shorter of the two). At some intermediate level, say a′b′, there must be a greater weight of air left above a′ than above b′, because pressure must decrease more slowly with height in warm air than in cold. As a result of this, there is a pressure gradient at the level a′b′ where none existed at the surface. Assuming freedom of movement between the two columns, air will move in obedience to this upper level pressure gradient, initially from a′ to b′ but eventually adjusted by the Coriolis deflection to a wind blowing into the paper. We call this a *thermal wind* because it results from horizontal temperature gradients.

The air flow in the middle and upper troposphere is largely dominated by the distribution of warm and cold air. From our northern hemisphere middle-latitude viewpoint we may say that warmer air on average lies to the south (column *A*) and colder air to the north (column *B*) so that the upper winds are predominantly westerly. As we go upward, there is a progressive change from the wind adjusted to the sea-level isobars to those obeying the thermal pattern. Hence the role of the latter in modern forecasting. The thermal winds are represented on charts of *thickness lines* (lines of equal thickness) which are in effect isotherms since the vertical thickness of an air layer between two pressure levels (e.g. 1000 and 500 mb) is a measure of its mean temperature: the warmer the air layer, the more it expands and the greater is its thickness.

Actual high-level pressures and winds (derived from radio-sonde and radar wind data) are represented by isobaric charts for various levels or, in British practice, by *contour charts* which show the varying height of a fixed pressure surface (e.g. 500 mb) just as relief contours show the varying height of the land. Contour charts show 'highs' (regions of warm air like column *A* in Fig. 7.3) and 'lows' (cold air like column *B*): the 500 mb surface is found at an average height in our latitudes of about 5·5 km (18 000 ft). Contour lines give a picture of the broad flow of air at the appropriate level and, like isobars, provide a measure of wind speed, so that contour charts have become essential analytical and forecasting tools.

Well before the Second World War there had been reports from time to time of high-flying aircraft being either slowed up or greatly speeded up or diverted far off course by unexpected and unpredictable winds of great velocity. Since the regular construction of high-level charts, we know that narrow belts of wind speed exceeding 100 m.p.h. (45 m/sec) and sometimes twice that value are a normal feature of the upper troposphere. These have been christened *jet streams*. Often only 2–300 km in width and with severe turbulence at their margins, jet streams are of the utmost significance to modern aviation – headwind jets being a nuisance but tailwind jets a boom – and their successful forecasting is therefore vital. There seems to be several types of jet stream: we return to this theme in later pages.

Pressure and Winds at Fronts

It is now possible to understand the notion of equilibrium at a front (Chapter 6) which maintains a sloping frontal surface and prevents the cold air from completely under-running the warm. If we think initially in terms of a stationary front and look back at Fig. 6.1(c), we see that there must indeed be a pressure gradient within the cold air directed towards the surface front but the resulting wind cannot escape the deflection due to earth rotation so that in fact it blows parallel to the front rather than across it. The isobars appropriate to this balanced condition must also lie parallel to the front and this is how we recognize a stationary front on a synoptic chart.

The argument is not disrupted if the front is moving, since an additional element of motion is added to both air masses. A moving front must be crossed by isobars, many of them if the 'push' on the front is strong, few if it is slow moving. Moving fronts lie along the axes of more or less pronounced troughs of low pressure and the isobars must change direction from one side of the front to the other: by convention they are always drawn more or less sharply angled or kinked across the front (see, for example, Figs. 6.4 and 6.5). The passage of a frontal trough is inseparable from pressure changes (again, more or less marked) and the wind changes obey in part the changed direction of the isobars.

With well-marked fronts the temperature contrast across the transition zone is found at all levels from near the ground to the very top of the troposphere. In fact the conditions implied in Fig. 7.3 are nowhere better realized than in a frontal situation, where a strong thermal wind must exist, blowing parallel to the frontal surface with the cold air to the left. This frontal thermal wind reaches its maximum strength just below the troposphere, where it constitutes the typical jet stream of middle latitudes. The position of the frontal jet is shown diagrammatically in Fig. 7.4, where the direction of the jet is out of the paper. These jet streams lie broadly parallel to the surface fronts (but displaced on the cold air side), move with them and, like them, may wave and become distorted. The most common is associated with the Polar Front and is called the *Polar Front Jet Stream* (PFJ). The frontal jet streams seem to have the vital function of rapidly evacuating air rising at the frontal surfaces. Other jet streams are found far from surface fronts – the most persistent being in sub-tropical latitudes – and have different origins.

The Mechanisms of Lows and Highs

The mechanisms of the various types of low and high-pressure systems are not always easily understood in detail but basically all are variations on relatively simple themes. Consider first the formation of a low in a region of previously uniform pressure. The key to the mechanism is the removal of air by some means or other from part of the region, a reduction of total weight of atmosphere which creates a surface pressure gradient, in response to which air must move in from all sides. Clearly, this is only the beginning of the story, for this motion of

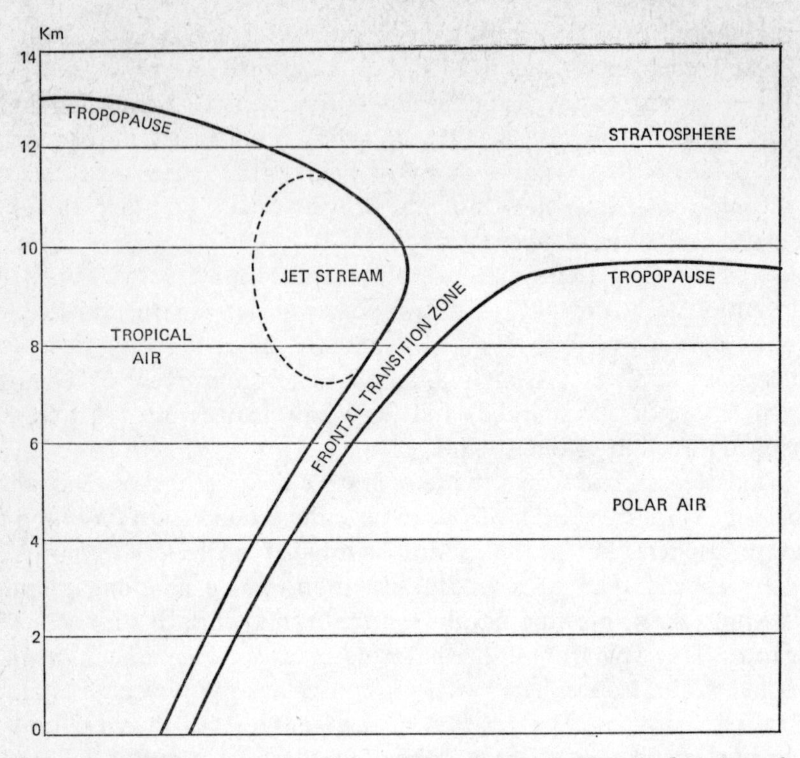

7.4 Diagrammatic vertical section across a front, showing the location of the frontal jet stream: the direction of the jet is out of the paper.

itself would soon even out the pressure deficit. If the low is to be maintained, air must be continuously evacuated. The initial direction of escape must be upwards – and here we recognize that the air flow at the base of a low is necessarily *convergent* (in the sense of the term introduced in connection with active fronts in Chapter 6) – but elsewhere must be outwards. In fact we are compelled to assume that outflowing winds at height, fed by the rising air (*divergence*) are a necessary part of the system and this high-level outflow must be at a rate exceeding that of the low-level inflow.

In the case of a high-pressure system it is easy to see that the same mechanism must operate but in reverse. Pressure rises over an area and there is a low-level outflow in deference to the newly-created pressure gradient. The loss is made good by descent of air or *subsidence* from above. To maintain the high, this low-level divergence must be more than compensated by air drawn in as a result of high-level convergence.

The horizontal movements entailed in these mechanisms are of course subjected to the Coriolis deflection. This ensures the characteristic low-level circulations – anticlockwise in the low, clockwise in the high – with a more or less complete reversal of direction at higher levels. A balanced condition is not to be expected while the systems are actively developing: a net outflow across the isobars must operate in the deepening low, a net inflow in the strengthening high. The systems may be kept going with little pressure change due to cross-isobar movement in the friction layer, the frictional inflow at the base of a low, the 'frictional leak' at the base of a high. In the dying stages, the net flow of air proceeds in the 'wrong' directions so that the low 'fills' or the high dissipates.

So it is that horizontal and vertical circulations are inextricably linked in these systems. Figure 7.5 illustrates these relationships and

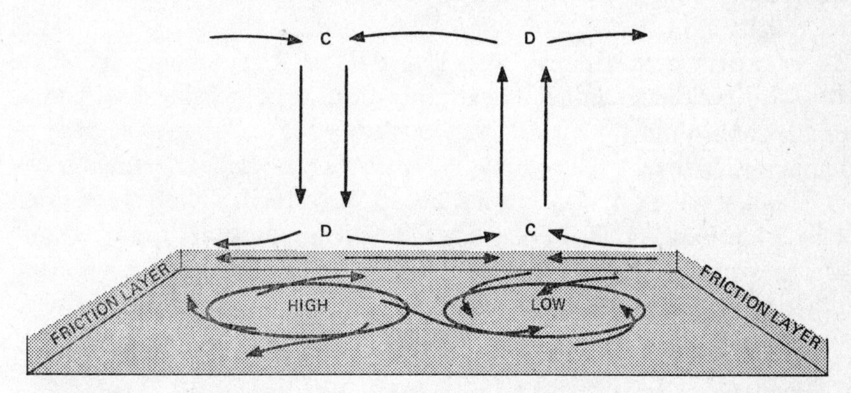

7.5 Horizontal and vertical circulations in adjacent high and low systems. C, convergence. D, divergence.

also shows how the low-level outflow from a developing high feeds the essential inflow to a neighbouring low, while a flow in the reverse direction occurs at high levels. It will also be noted that enormous weights of air are shifted to and fro in these systems at various levels but what our surface barometers record are the *net* changes which are small in comparison.

We have been concerned in this section with certain necessary accompaniments of the formation of pressure systems, but not yet with their causes, with the 'how' rather than the 'why'. The causes of these developments are not completely understood but it does seem that the mechanisms can be set in motion at any point, so that if, for any reason, there is low-level convergence, a surface depression will develop, but equally a high-level convergence will result in the formation of a surface high. We can tentatively recognize the origins of pressure systems as either thermal (to do with the heating or cooling of the atmosphere near the ground) or related to convergences or divergences at upper tropospheric levels, or to accumulations and deficits of air at low levels due to purely orographic causes, or to more than one of these in combination.

Whatever their origins, all low-pressure systems share the common characteristic of ascending air, whether this is localized in distinct belts as in frontal lows or general as in other types, to which they owe their associated cloud and bad weather. On the other hand, the key to the understanding of anticyclonic weather is the accompanying subsidence, in which air is warmed adiabatically and therefore effectively dried. It is to this fact that anticyclones owe their reputation for good weather, although this is not always deserved: our summer highs can be relied upon for fine or fair weather (Plate 13) but winter highs developed in cold air may well give dull or foggy conditions (Plate 14). Anticyclonic weather is certainly dry, since subsidence imposes great stability on the air and this is clearly visible on a tephigram or temperature–height diagram as a *subsidence* or *anticyclonic inversion* at heights sometimes as low as 500 m (say 1600 ft) but more often 1 or 2 km (see Fig. 7.6). Very dry air (wide separation of temperature and dew-point curves) distinguishes the anticyclonic inversion from other types.

Thermal Systems

These are caused by heating or cooling of parts of the earth's surface. *Thermal depressions* or heat lows may occur over islands and peninsulas on warm summer days, due to the different properties of land and water (Chapter 3). With strong insolation the land surface becomes warmer than the surrounding sea and a shallow air layer above the land, warmed in turn, expands vertically. Towards the top of this layer (at a height of the order of a kilometre), high pressure develops over the land area as air accumulates from below: a pressure gradient now exists at this level, in response to which air flows outwards thus reducing the total weight of atmosphere above the land and increasing it above the surrounding sea. The surface low is now established, the essential mechanism is set in operation and descending air over the sea and inflowing winds at low levels complete the circulation.

Essentially the same *thermal circulation* occurs on a smaller scale in and around any convection cloud or, for that matter, in a heated

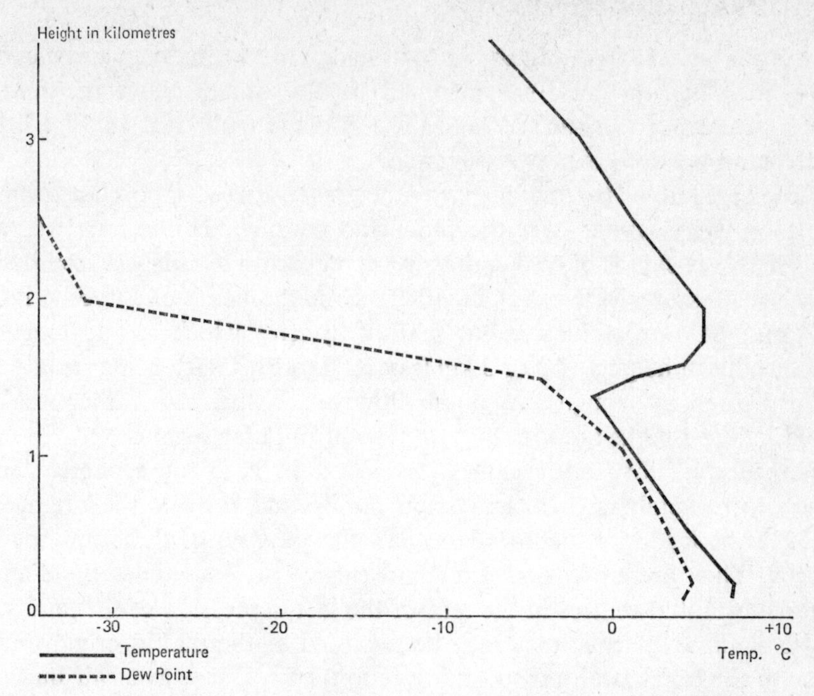

7.6 A subsidence or anticyclonic inversion on the temperature-height diagram.

kettle of water or a convectively warmed room. Looking forward to Chapter 9, we recognize that this is also the explanation of the origin of sea breezes. On a continental scale, the same mechanism operates to give what is known as a *monsoon low*, which dominates the summer weather and climate over vast areas. It also appears to be, as we shall see, one of the mainsprings of the global circulation.

The commonest thermal depressions we experience are the shallow *Polar Air Lows* that arrive in winter embedded in a cold Pm airstream. The constant heat supply from relatively warm underlying waters must be important in their origin but the latent heat released during condensation in frequent cloud formation no doubt acts as an additional energy boost (as it must in any depression). The shallow thundery lows of summer that reach us from the south are of similar character. Most thermal lows give an intensification of showers or afternoon thunderstorms rather than continuous bad weather, but in stable air they may be quite innocuous.

The exact reverse of the above mechanism gives rise to a *thermal high*. The cooling of part of the surface is communicated to the overlying air layer which contracts or shrinks vertically, leaving a deficit of air in its higher levels (an upper low) which in turn induces an inflow. This increases the total weight of air in the column above the land, thus producing the surface high. The process embodied in Fig. 7.5 (left-hand portion) is now set in train.

The best-known thermal highs are the great *seasonal* anticyclones that develop every winter over the snow-clad continental interiors, where under vast cushions of cold dense air are registered the highest recorded surface pressures (for example, 1080 mb in the heart of the Siberian High). These are extensive but shallow systems which are no longer discernible at heights of 2 or 3 km (say 6000 to 10 000 ft), being replaced above often by troughs. Again shallow but in this case rather weak cold-induced highs are usually to be found in Polar regions.

More transitory anticyclones developed in cold air separate the successive members of a depression family and provide the brighter interludes, lasting sometimes less than than a day, of these unsettled spells. They are often no more than ridges but sometimes build up into substantial centres in the rear of the last depression of the family. Also shallow systems, they may be regarded as thermal in origin, but due to the horizontal transport (advection) of colder heavier air into a given area rather than to the consequences of surface cooling in situ.

All the anticyclones described so far are known as *cold highs* and it is easy to see that the extra weight of air is due to the presence of a cold dense layer in the lower troposphere. There are other highs in which temperatures are often warmer than normal throughout the lower and middle tropospheres and the excess weight must be attributed to relatively cold air in the upper troposphere which follows from the unusually high tropopause associated with these systems. Since pressure decreases with height only slowly in warm air, these highs, unlike the shallow cold systems, exist throughout much of the troposphere, showing up strongly on the 500 mb contour charts, and are the most stable, slow-moving and persistent of pressure systems. Although known as *warm highs,* these are not to be explained in simple thermal terms. The *Sub-Tropical Highs* are of this category (the Azores High being our nearest example) and are integral parts of the general circulation. From the Azores High, temporary warm anticyclones sometimes extend towards Britain either as ridges or as separate systems that bud off the main centre, giving us the best of Tm weather. Stationary cold highs may gradually change into warm types through the agency of the subsidence process.

The Effect of High-Level Convergences and Divergences

The travelling lows and highs of middle-latitude synoptic charts seem in large part to be responses to upper tropospheric developments in the form of accelerations or retardations of air flow or cross-isobar (ageostrophic) movements which create convergences and divergences at these levels and trigger off the mechanisms depicted in Fig. 7.5. For example, the middle and upper tropospheric flow in these latitudes is often along curved paths, as shown on 500 mb or 300 mb contour charts, and curved flow, as we have seen, entails a departure from geostrophic conditions. In the axis of a trough (cyclonic curvature) the wind has less than the geostrophic value (see Fig. 7.1) but in the region of straight flow ahead the wind increases to geostrophic. This is an acceleration and, since more air leaves a given area than enters it, a high-level divergence results, which will generate a low-level convergence beneath it. Thus the area ahead of the axis of an upper trough is a favoured zone for cyclonic development. Behind a trough or ahead of a ridge, where the air slows down from supergeostrophic to geostrophic speeds, there is a high-level convergence, making for anticyclonic development as seen on the surface chart (Fig. 7.7(a)).

Discontinuous jet streams of the frontal variety are also a feature of upper tropospheric flow in middle latitudes. At what is called the jet entrance (Fig. 7.7(b)) the pressure gradient tightens so rapidly that the

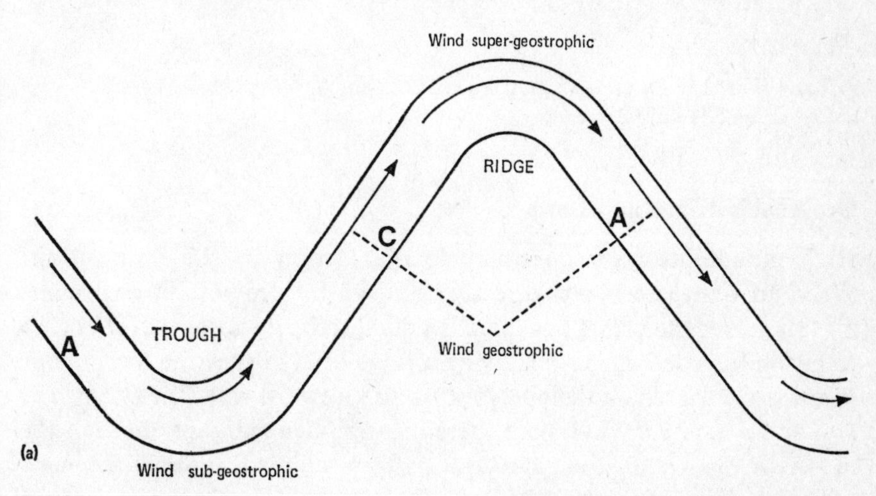

7.7(a) Contour pattern showing areas favouring cyclonic development (C) and anticyclonic development (A) at low levels. The length of the arrows broadly denotes the wind speeds.

wind is out of adjustment, having less than the geostrophic value (subgeostrophic). As a result, air is here transported across the isobars (or contours) from high pressure (warm air) to low pressure (cold air), giving a high-level convergence which will tend to generate a surface divergence (anticyclonic development) below it left of the entrance, while leaving a high-level divergence (and therefore surface cyclonic development) to the right of the entrance. Conversely, at the jet exit, where the pressure gradient slackens and the associated wind is supergeostrophic, air is thrown to the right, where the upper convergence (and surface anticyclone) are to be found. Employing a variant of Buys Ballot's Law, we could say that if we stand at either jet entrance or exit with our back to the tightest gradient, then the surface cyclonic development (i.e. bad weather region) is to our left: these are the areas marked *C* in Fig. 7.7(b).

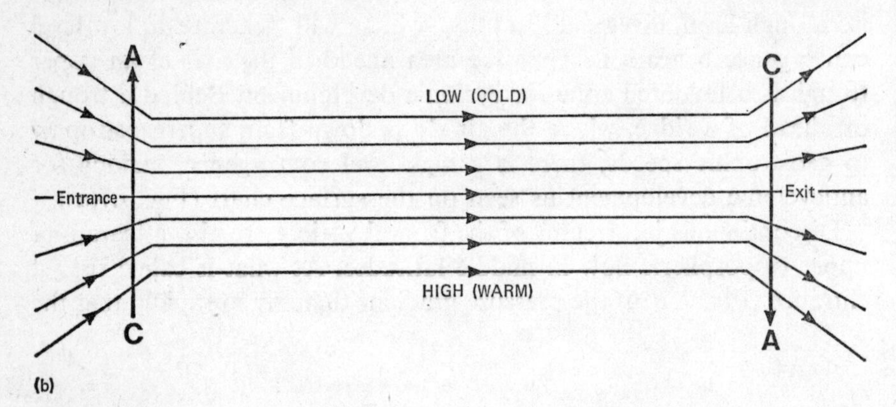

7.7(*b*) *Westerly jet stream showing areas of low-level cyclonic* (*C*) *and anticyclonic* (*A*) *development.*

Orographic Highs and Lows

High mountains offer considerable obstruction to the flow of air. When an airstream moves directly against a high mountain wall, some air flows over the crests but there is bound to be a substantial piling up of air on the windward side, giving a ridge of high pressure (*orographic high*), while there is a deficit of air on the leeward side (the *orographic low* or *lee depression*). Where there are well-defined gaps through the mountain barriers, the pressure pattern created ensures strongly funnelled gap winds, of which the classic example is the Mistral of the Rhône valley in France which blows with great force and persistence between an orographic high north of the Alps and a lee depression over the north Italian plain. This type of low is primarily non-frontal but fronts may be drawn into its circulation. This frequently occurs over northern Italy in the Mistral situation when a cold front heralding Polar Maritime or Arctic air, which is elsewhere held up against the northern flanks of the Alps and Central Massif of France, finds itself accelerated through the Rhône valley and incorporated into the so-called 'Mistral Low'.

Tropical Low Pressure Systems

The intense tropical lows which are variously known as *hurricanes* in the West Indies, *cyclones* in the Indian Ocean, *typhoons* in the China Sea and *Willy-Willies* in Northern Australia, are not quite in any category we have yet discussed. They appear on synoptic charts (Fig. 7.8) as exceptionally deep lows with a concentric pattern of tightly packed isobars. The central pressure is generally around 950 mb but may be much lower: the lowest recorded pressure – 887 mb – was measured in the heart of a Pacific cyclone. The systems may be 80 or

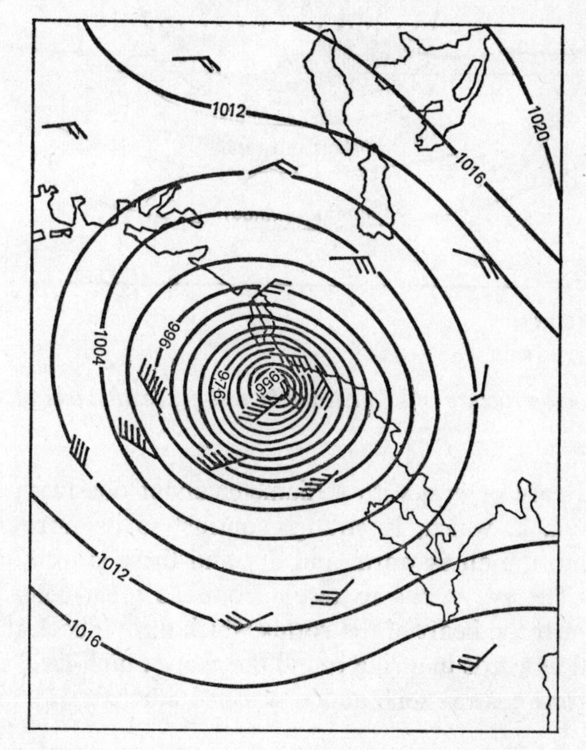

7.8 Hurricane 'Edna', 11 September, 1954, 1830 hr GMT.

800 km (50 or 500 miles) in diameter and wind speeds often exceed 100 m.p.h. (about 45 m/sec).

Until recently, comparatively little was known of these violent storms and this is not surprising since Nature has probably conceived nothing quite as frightening and the interests of anyone caught up with them, on land, sea or in the air, are centred on survival rather than careful weather observation. However, determined investigation – involving aircraft and aerological sounding as well as surface reports and, lately, satellite photography – has enabled us to construct something of a picture of the typical hurricane. A simplified idea of the supposed structure is given in cross-section and plan in Fig. 7.9. There

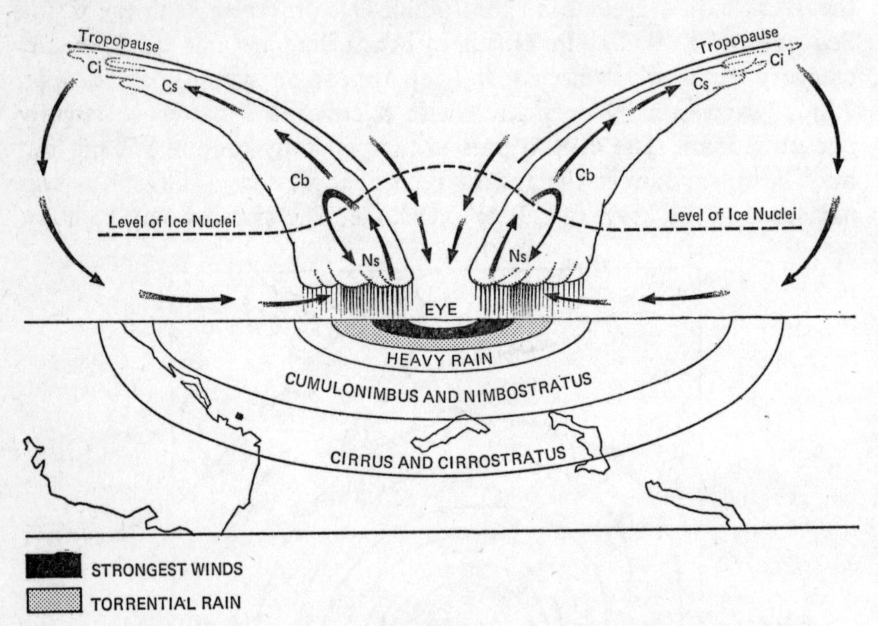

7.9 Suggested structure of a hurricane in vertical section and plan (after Bergeron).

is a central core or *eye*, with a diameter about one tenth that of the whole system, in which, in strange contrast to its surroundings, the wind is calm, the air is warm and dry and there is little or no cloud (Plate 15). The eye seems to result from the great centrifugal force developed near the heart of the vortex which throws back the inflowing air to spiral upwards in a ring round the centre: high-level air is drawn down into this central zone and is warmed adiabatically and dried in descent.

Around the eye is a great cylinder of cloud, which in the case of the

Pacific typhoon 'Marge', illustrated in Plate 15, exceeded 35 000 feet (nearly 11 km) in height. The strongest winds blow beneath the inner part of this cloud ring, within which cumulonimbus towers, seething with fierce updraughts and downdraughts, are embedded. Here the Bergeron mechanism operates vigorously to give torrential rain falling at a rate of an inch (25 mm) or more an hour. Satellite pictures reveal a spiral arrangement of the thickest clouds around the centre. To an observer, the storm seems to blow from a particular direction, then follows the deceptive calm of the eye and then the storm 'returns' in full fury, with the wind now in the opposite quarter.

In origin, hurricanes appear to be in part thermal systems, since they form only over the warmest tropical waters, with surface temperatures exceeding 27°C (80°F). Much of the energy that drives them comes from the latent heat freed during abundant condensation in air kept highly charged with moisture evaporated from the warm underlying water. They form within the Tropics but not nearer than 5° from the Equator (where the Coriolis deflection is too small for a circulation to develop). It was once thought that they formed on or near the Inter-Tropical Front (now an out-moded concept): at any rate general low pressure, i.e. tendency to convergence, seems a necessary condition for their formation and this is often provided by shallow non-frontal troughs (easterly waves) which are commonly found in the trade wind flow of these latitudes (see Chapter 8).

Hurricanes occur mainly in late summer and autumn. They move quite slowly, generally less than 15 m.p.h. (7 m/sec), in compliance with the general circulation of tropical regions, that is, westwards at first, then curving polewards towards middle latitudes where they turn eastwards (see Fig. 7.10) and soon die out over cooler waters. Some, however, draw fronts into their circulations and become revitalized as middle-latitude depressions: more than one of our own homely frontal lows began life as a West Indies hurricane. Sometimes the storms move over land where, cut off from the inexhaustible supply of heat energy provided by the warm ocean and their winds countered by greater surface friction, they usually degenerate rapidly. It is thus in coastal areas that the great hurricane catastrophes usually occur (Chapter 12) but these systems occasionally move vigorously inland at exceptional speeds like the New England hurricane of 1938 which whipped 150 miles (240 km) inland at 65 m.p.h. (29 m/sec) and hurricane 'Hazel' which left a wake of devastation from the Caribbean to the Arctic in the notorious hurricane season of 1954.

Tornadoes are even more violent than hurricanes but being much smaller (100–200 m in diameter) their highly destructive effects are much more localized. They form over land in unstable damp air in strongly convectional situations during afternoon or early evening in spring or summer. They are again vortical systems with an intense cyclonic spin around a usually inclined axis. The wind speeds are quite fantastic, reaching perhaps 200–300 m.p.h.: these can only be estimates, based on the damage caused. Of uncertain origin, tornadoes are always associated with large active cumulonimbus clouds: it has been suggested that an interaction between the back-spreading down-draught and the inflowing current towards the rear of the cloud may be involved in the initiation of the cyclonic whirl. There is always a characteristic *pendant* or *funnel cloud* extending from the cumulonimbus base towards the ground (Plate 16). This is due to the great pressure

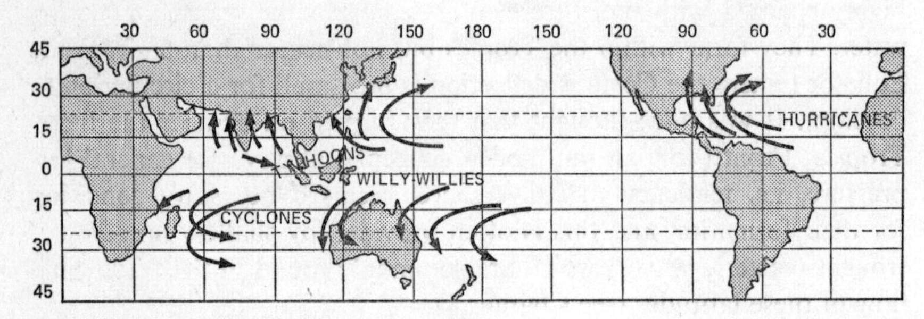

7.10 Frequent tracks of tropical hurricanes.

fall in the centre of the whirl which causes adiabatic cooling and, hence, condensation. Some observers claim to have seen a narrow eye within the funnel.

The tornado arrives to the accompaniment of violent thunderstorms, heavy rain, sometimes giant hailstones and a general roaring sound compounded of wind, thunder and the noise of destruction. The damage results from the combination of pressure and suction effects exerted by the wind: dust and debris (which may include trees, roofs and vehicles) are sucked skywards and flung aside, thus becoming powerful tools of destruction. Tornadoes travel usually at 30 to 40 m.p.h. but occasionally much faster, in paths that may be long or short, straight or tortuous. Their duration at any one spot will not be long, but within a narrow swathe devastation may be complete.

North America suffers most from these 'twisters' of which, on average, about 150 are reported annually, mainly in the Mississippi lowlands. Temperate regions have occasional visitations of rather less virulent examples: a few are reported in Britain most years. The equivalent of the tornado over the sea is the *waterspout*, frequently seen over warm waters, though sometimes these are land tornadoes that have moved out to sea. Some gentler versions – *fair weather waterspouts* – are more akin to dust-devils (Chapter 4).

Requirements of a General Circulation

And so to the major problem in meteorology, the grand scheme in which all the elements we have considered – the air masses, the fronts, the pressure systems – find their due place. It must be said at once that, despite voluminous writings and a succession of 'models', a complete and satisfactory explanation of the general circulation still eludes us. What is not in doubt is the need for a general circulation of a certain kind. In Chapter 3 (page 53) we drew attention to the unequal heat endowment of the earth's surface which can be evened out only by the transport of excess heat from low to high latitudes. Later in that chapter (page 63), we saw that precipitation/evaporation relationships in different latitudes demand a transport of moisture both polewards and equatorwards from the sub-tropics. Yet another consideration has not yet been mentioned: the earth rotates from west to east, therefore easterly winds by virtue of their frictional drag would tend to slow down the rate of rotation, while westerly winds would act to accelerate it. In regions of easterly winds, momentum (strictly, angular momentum) is transferred from faster-moving earth to slower-moving air, while in westerly wind regions a reverse transfer occurs, from faster-moving air to slower-moving earth.

We need not doubt that the atmosphere successfully solves these problems at least in the long term. Low latitudes are not progressively warming up, nor high latitudes cooling off. Tropical and temperate belts are kept supplied with moisture. The earth continues to rotate at a constant rate, which means that the areas under easterly and westerly winds are about equal and that momentum is constantly being transported from the former to the latter. The problems are those of our understanding of these complex relationships and processes.

Any viable scheme of the general circulation must accord with observed facts. Our knowledge of the mean surface pattern of pressure and winds has been built up over many years and is often generalized into the *idealized planetary circulation* shown in Fig. 7.11. This is a hypothetical circulation referring to a uniform globe and is far from the realities of the situation, approaching them perhaps only over the wide expanses of the Pacific Ocean, but it will serve as a stepping stone to fuller understanding. It shows the dominant wind belts of the globe, the *westerlies* of middle latitudes, flanked to poleward by the relatively narrow belt of *polar easterlies* and to equatorward by a broad band of low-latitude easterlies better known as the *trade winds*.

The pressure distribution is related to these winds belts in accordance with Buys Ballot's Law. In terms of the northern hemisphere, the trades blow between a sub-tropical belt of high pressure around latitude 30° (the 'Horse Latitudes') and an equatorial low-pressure region (the 'Doldrums'): because of surface friction, they blow somewhat across the mean isobars towards low pressure as 'north-east trades'. The westerlies become more like south-westerlies, blowing between the sub-tropical high and a low-pressure zone ('sub-polar low') at 55°–65°N, while the polar easterlies (or north-easterlies) blow out from the polar high.

More recent observations have filled in the picture in the vertical and the right-hand portion of Fig. 7.11 depicts broadly in profile the region of mainly westerly and easterly winds (the latter shaded) in the troposphere. This shows that the westerlies, which reach the surface

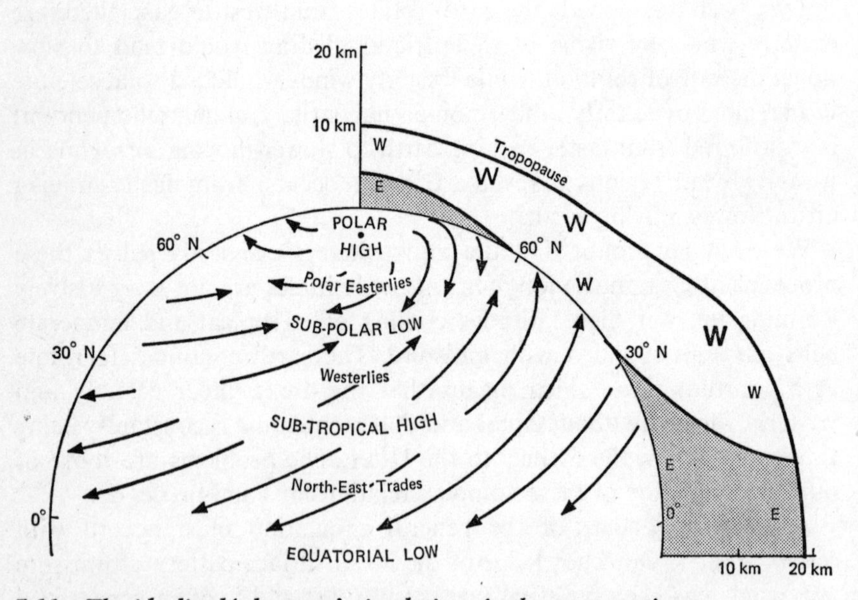

7.11 The idealized 'planetary' circulation: in the cross-section, easterly wind belts are shaded and the size of the letters E (easterlies) and W (westerlies) broadly indicates the wind strength.

only in middle latitudes, are much more extensive in the middle and upper troposphere. The polar easterlies are represented only by a shallow capping 2 or 3 km (6–10 000 ft) in depth (this varying somewhat with the season). The low-latitude easterlies, widespread at the surface, contract with elevation but are still found at all heights within 10° to 20° of the equator. The dominant westerlies strengthen with height up to about 12 km (40 000 ft) and reach maximum strength above about latitude 30° in a persistent feature known as the *sub-tropical jet stream* (STJ).

We see from Fig. 7.11 that the pattern of air flow in the middle and upper troposphere is simpler than that near the surface. Another view of this is provided by Fig. 7.12, which shows on a polar projection for the northern hemisphere, mean 500 mb contour charts for January and July: it will be remembered that these represent the flow pattern at 5–6 km (say 17 000–20 000 ft). The dominance of the westerlies is strikingly confirmed as a vast *circumpolar vortex,* occupying all latitudes except near the equator, where easterlies take over. These upper westerlies blow between high contour values towards the equator and low over the pole. The pattern of contours is near (but significantly not quite) circular and comparison of the two charts shows that the circumpolar whirl is both stronger and more extensive in winter. Very tight gradients in places reveal the sub-tropical jet stream (which however shows at its best on 200 mb charts): the polar front jet stream (page 149) is too variable a feature to stand out on these average charts.

Early Circulation Models

The idea of an essentially thermal circulation driven by the temperature difference between low and high latitudes goes back to 1686 and Edmund Halley (better known for his comet). On a stationary globe this would result in a northerly flow at the earth's surface and a return flow of southerlies at higher levels (in the northern hemisphere), the two linked by rising air over equatorial latitudes and descending near the pole. In 1735, George Hadley amended this scheme to allow for the effects of earth rotation, thus achieving upper westerlies and surface easterlies. Although overall surface easterlies are geophysically impossible, what came to be known as the 'Hadley cell' seemed reasonably to explain the circulation in low latitudes and remains part of most circulation schemes to this day.

From about the mid-nineteenth century, with more awareness of friction effects and the benefits of a fuller picture of surface pressure and wind distributions, further advances were made by Ferrel, Maury and others. The Hadley cell was accepted for the tropical and sub-tropical regions. The, by now, recognized sub-tropical high pressure belt was seen as a zone of descending air feeding not only the return flow towards the equator (trades) but also the middle-latitude westerlies (sometimes known as the Ferrel westerlies). Later it was suggested that the high-latitude circulation was also a thermal cell, driven by the excess cold of the polar cap: in this scheme, air descends in the (thermal) polar highs, flows outwards (polar easterlies) to ascend in the sub-polar lows and return as upper westerlies. All of this fitted well with observed

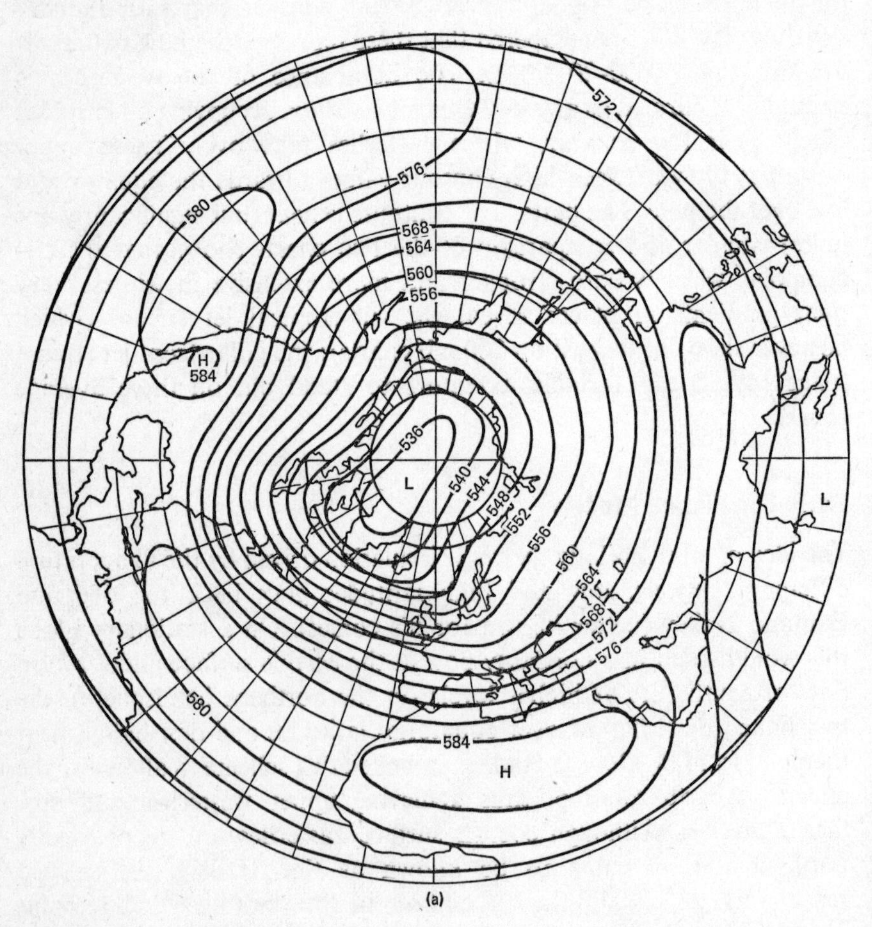

7.12 The mean flow at 5–6 km as shown by mean 500 mb contours, northern hemisphere (a) July (b) January (after Scherhag).

facts. There were problems in middle latitudes but arguments were clearly tending towards a three-cell model of the mean circulation.

The Norwegians were able to fit their air-mass and frontal concepts into this scheme. The two great trade wind streams converged in the vicinity of the equator at the Inter-tropical Front. The warm westerlies (Tropical air) rose above the cold polar easterlies (Polar air) at the Polar Front. The most complete statement of the three-cell model came from the Swedish meteorologist Rossby in 1941 (Fig. 7.13): he called the two thermal cells already described the 'Trade Wind' (=Hadley) cell and Polar Front cell respectively, with a 'Middle' cell between them. It can be seen from the diagram that the upper limb of

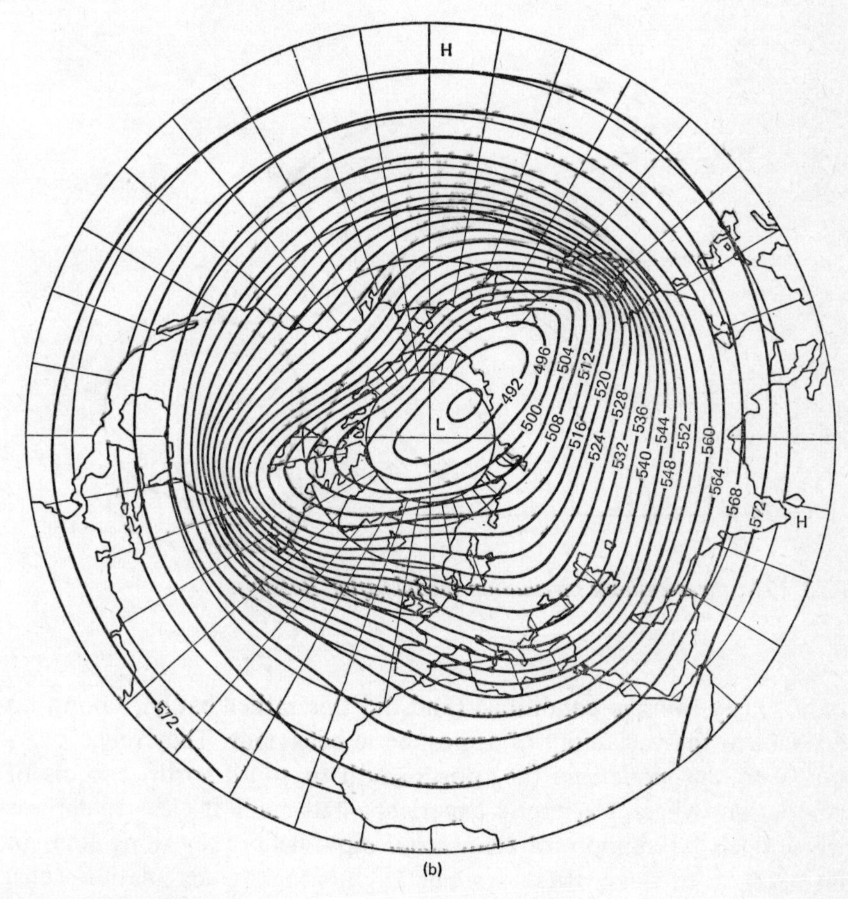

7.12 *See p. 164*

the Middle cell gives the 'wrong' answer, i.e. a northerly flow affected by the Coriolis deflection gives easterlies whereas the winds actually observed here are predominantly westerly. Rossby explained this anomaly as a frictionally enforced westerly flow imposed by lateral mixing with the upper limbs of the cells on either side.

Eight years later Rossby largely abandoned his scheme not only because of the problematic middle cell but also in the face of other objections which he himself recognized. Most important perhaps was the realization, probably inevitable at this time and state of knowledge, that there might be no such thing as a single steady-state circulation, applying along all lines of longitude and at all times, that could be adequately depicted in a diagram such as Fig. 7.13. Such schemes could

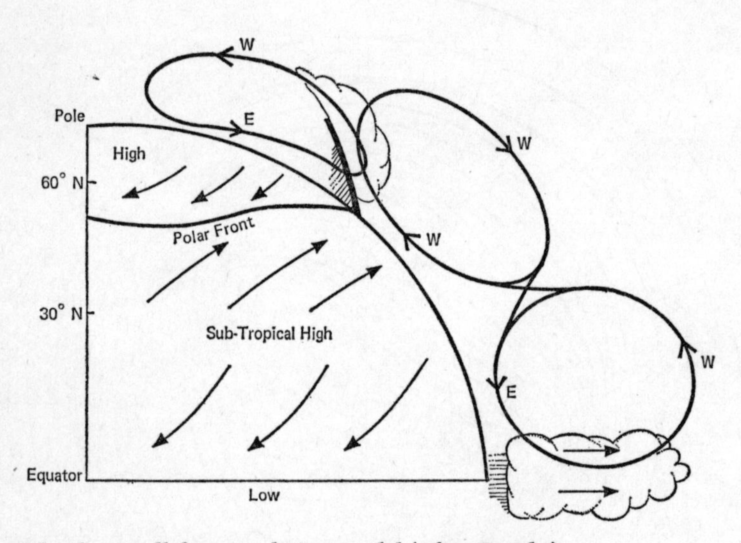

7.13 The three-cellular circulation model (after Rossby).

only satisfy average conditions (and did this rather badly), taking no account of the variability of atmospheric behaviour. They relied overmuch on the *meridional* (i.e. north-south or south-north) aspects of circulation, whereas synoptic experience had made it clear that these were much less important than *zonal* movements (i.e. along lines of latitude). Moreover, these systems of interlocking circulation cells, however tidy, did not necessarily engineer the required transports of heat and momentum between latitudes. Attention shifted to synoptic charts rather than to average-condition maps, to the dynamic rather than the static picture, to what had earlier been regarded as 'noise' rather than 'signal'.

Realities in the General Circulation

The initial step away from the highly idealized surface circulation (Fig. 7.11) is to acknowledge the existence of the continents and oceans. Their effect broadly is to break up the pressure and wind belts into separate centres. In the northern winter the intense cold of the Eurasian and North American land masses gives rise to two great thermal highs which dominate the circulation of the northern hemisphere. They extend northwards to split the sub-polar low into the Icelandic Low (North Atlantic) and the Aleutian Low (North Pacific). The sub-tropical anticyclones over the oceans – Azores and North Pacific Highs – are overshadowed in strength by the great continental highs.

In the northern summer the thermal systems over the continents are transformed. A great monsoon low develops over southern Asia and a smaller one over North America. These completely interrupt the sub-tropical high-pressure belt, leaving only the Azores and North Pacific anticyclones, which however stand out more strongly than in winter. The Asiatic monsoon low is so deep that it swallows up the equatorial low in the Indian Ocean, creating a continuous pressure gradient from the southern hemisphere sub-tropical high to the Asiatic low. The sub-polar lows are weak in summer, overshadowed by the continental monsoon lows.

In the southern hemisphere where there is much less land, the mean pressure distribution is less cellular and more like the idealized pattern on a uniform globe. Nevertheless the sub-tropical high still consists of three centres, one in each ocean, especially in the southern summer. But because of the virtual absence of land in the latitude concerned, the sub-polar low extends as an almost unbroken belt.

Figure 7.14 helps us to see in a general way how the global surface patterns of pressure, air masses and fronts are interrelated. It shows, for January and July respectively, the average sea-level isobars and air flow, on which are superimposed the location of the air masses and the average positions of the major fronts. The high pressure centres are air-mass source regions and the lows coincide broadly with the frontal zones. It is the location of the pressure centres that determines where contrasted air masses shall converge (frontogenesis) or diverge (frontolysis). Thus Pm air sweeping south-eastwards in the circulation of the Icelandic Low confronts Tm from the Azores High at the Atlantic Polar Front. But further east the air flow is on average divergent and frontogenesis does not occur. To some extent, the familiar wind belts may be re-named, or given additional names in air-mass terms but, as Fig. 7.14 shows, air from diverse sources makes up the 'trades' and the 'westerlies' are partly polar, partly tropical, in origin.

Although we may learn a good deal from such maps, we must also bear in mind that averages in meteorology often conceal as much as they reveal. The pressure centres are not fixed features, the two polar highs in particular being weak and variable, while even the 'permanent' sub-tropical highs fluctuate in strength and location not only seasonally but over much shorter time periods. The so-called 'Icelandic Low' does not appear as such on any individual synoptic chart: it is merely an abstraction representing the fact that travelling depressions are particularly frequent in the Icelandic region. The major fronts, lines on the mean circulation maps, are actually pulsating zones of cyclogenesis. These maps have their uses but they provide a static view (an abstract one at that) when we really require a moving picture, a cinematic view. For fuller understanding, we must move in this direction.

The mean 500 mb contour charts of Fig. 7.12 show a close approach to a purely zonal flow, a near-perfect circumpolar whirl. Individual charts, however, reveal that the flow is far from regular but instead undulates in a series of ridges and troughs around an amoeboid-shaped and ever-changing polar low. These undulations are known as *upper long waves* or *Rossby waves* (after their discoverer). Figure 7.15 is not an actual chart but shows a common 500 mb contour pattern which gives an idea of the number and dimensions of the waves. Embedded within these meandering upper westerlies are belts of particularly strong winds (i.e. close-packed contours) – the Polar Front Jet Stream.

Figure 7.15 also shows, broadly, how middle tropospheric and surface patterns are related, since a typical system of surface fronts has been inserted. It will be remembered that high contour values represent warm air and low values cold air. The troughs in the pattern are therefore regions of cold polar air, tongueing equatorwards (the shading helps to draw attention to these) and the ridges are regions where tropical air is displaced polewards. The waving upper westerlies are considered to be one of the basic features of tropospheric behaviour and here we see the reason for the ceaseless tussle between air masses at

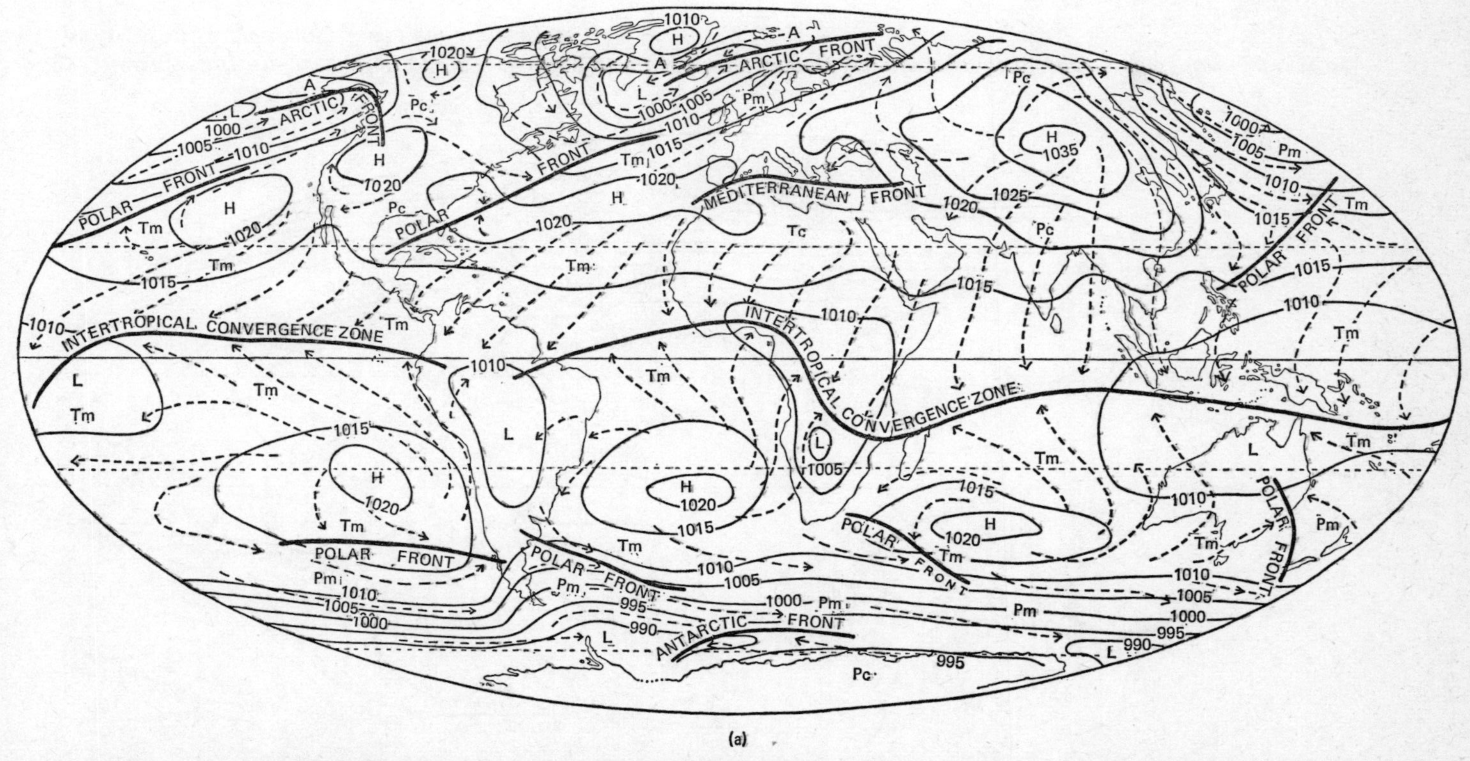

7.14(a) Average surface circulation in January, showing mean sea-level isobars (continuous lines), mean winds (broken arrows), mean positions of fronts (heavy lines) and air masses.

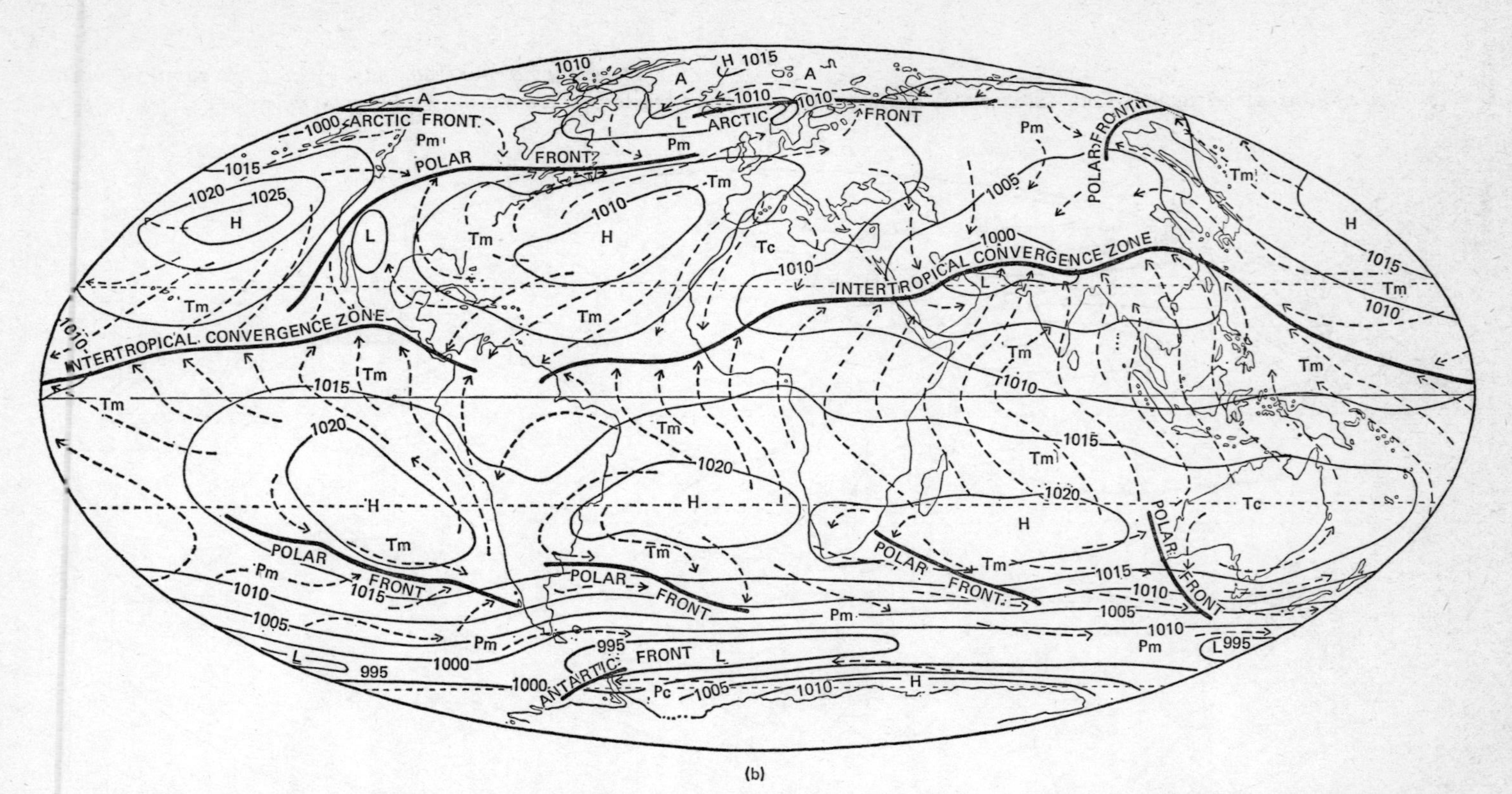

7.14(b) *Average surface circulation in July, showing mean sea-level isobars (continuous lines), mean winds (broken arrows), mean positions of fronts (heavy lines) and air masses.*

the Polar Front. Cyclonic activity, as we saw earlier in this chapter, occurs characteristically under the east side of these troughs and anticyclones tend to develop under the west side. An upper wave is linked sometimes to one large surface depression but more often, as the diagram shows, to a depression family.

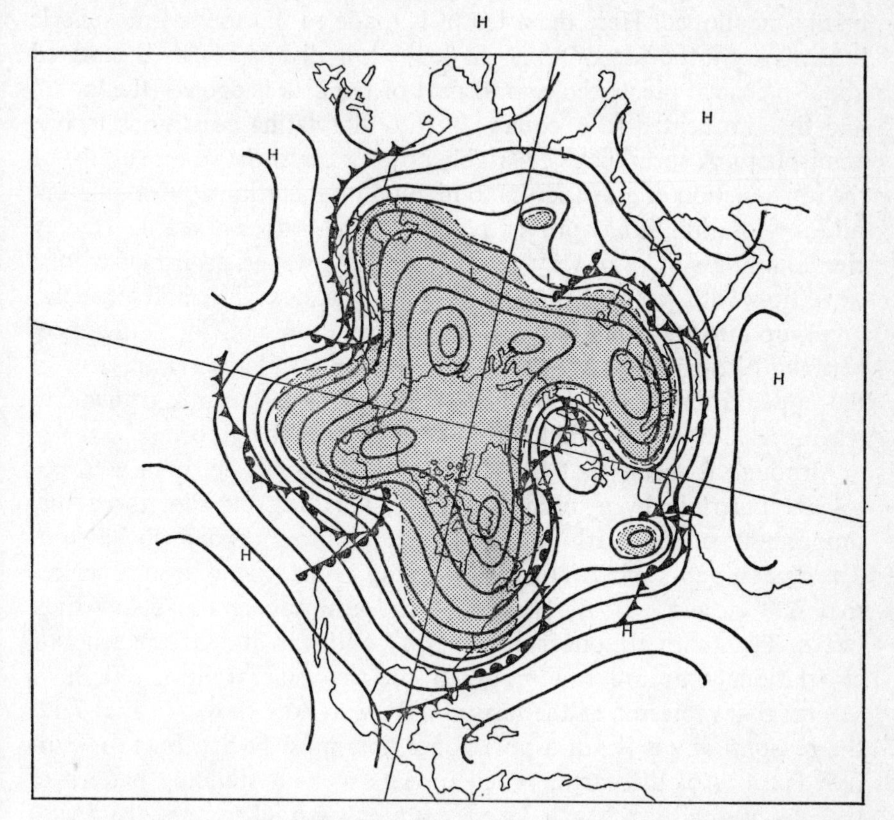

7.15 Schematic circumpolar chart showing long waves in the upper westerlies (after Palmén). Thin lines are 500 mb contours; the broken line represents the Polar Front at that level; the extent of polar air at 500 mb is stippled: surface fronts are conventionally shown.

The waves move round from west to east, taking their attendant sequence of surface depressions and anticyclones with them, in an uneven, pulsating and constantly distorting pattern. Sometimes a particular trough becomes so distorted that the equatorward portion becomes pinched off, remaining as a *cut-off low* or *cold pool* of air stranded in warmer latitudes. Part of an upper ridge may similarly become detached. Examples of both are shown in the diagram.

Why should the upper tropospheric flow be disturbed in this way? It seems on the basis of both mathematical and experimental models developed mainly in the last two decades, that we must accept this mode of behaviour as inevitable, the properties of the atmosphere being what they are. The remarkable 'dishpan' experiments must be briefly mentioned. Here the attempt is made to represent atmospheric behaviour by the use of water in a shallow dish that can be rotated about a central pivot: the central part of the dish is cooled (the 'pole') and the rim heated (the 'equator'), thus simulating conditions in one hemisphere. A sprinkling of suitable powder on to the water surface or the introduction of a dye serves to highlight the motions. With different rates of rotation, it has proved possible to reproduce a simple Hadley circulation, a system of very steady open long waves, an irregular long wave flow with jets and vortices forming and decaying, and a general break-up into small convection cells. Despite some obvious differences between dish-pan and atmosphere, a pattern strikingly reminiscent of the upper tropospheric flow is achieved at a rotation rate equivalent, when scaled down, to middle-latitude speeds on the globe.

Although it seems that the Rossby waves would occur in a troposphere underlain by a uniform surface, it is unlikely that the actual topography of the earth is without effect on atmospheric behaviour. The great mountain barriers of America and Asia extend to such heights that half or more of the weight of the atmosphere lies below their peaks. The thermal contrasts between continent and ocean generate circulations that are superimposed on the general flow. We have several times referred to the near-zonal mean flow shown in Fig. 7.12: the reasons why it is not a pure zonal flow must be attributed to surface features of the earth. What appears to be a standing pattern of troughs over eastern North America and eastern Siberia may be due to an 'anchoring' effect of the Rockies and Himalayas, perhaps reinforced by continental effects (since the features are much stronger in winter). The weaker trough over eastern Europe remains puzzling, which, together with other complexities, shows that there is more to be learned about this aspect of the circulation.

A Realistic Circulation Model

What, then, remains of the notion of a general circulation, amid this welter of detail? It was in fact back in the 1920s when most workers were still concerned with the various cog-wheels of a mean meridional flow, that Jeffreys, perceiving that vital clues might lie rather in the daily disturbances, suggested that the primary requirements of the circulation – the poleward transport of heat and momentum – could be satisfied by the travelling disturbances of middle latitudes. We are familiar with the idea of turbulence as a mixing process in the vertical sense. The new approach was to see the travelling lows and highs and their counterparts in the upper tropospheric waves as large-scale turbulent eddies in the horizontal field, persistently shifting warmth and momentum polewards. A continuous undisturbed Polar Front would clearly act as a barrier prohibiting air-mass encroachment but the disturbances act as travelling exchange mechanisms spraying warm air polewards and cold air equatorwards.

Following two decades of neglect, the hypothesis was revived with the availability of more upper-air data after World War Two and mathematical calculations demonstrated its feasibility. There appears to be no need for a global mean meridional circulation: a very large part of the required transfers can be explained by the horizontal eddies of middle latitudes.

Further attention to synoptic detail confirms and clarifies the key rôle of these systems. Sometimes the cold anticyclone in the rear of the last member of a depression family builds up strongly and coalesces with the sub-tropical high to equatorward: with general subsidence the intervening front is soon wiped out by frontolysis and there is no longer a barrier between the air masses. In this way polar air gains entry into lower-latitude circulations.

In fact the general circulation presents a constantly changing picture and synoptic experience suggests that the fluctuations are not haphazard but show a swing between two extreme patterns. In one the circumpolar vortex is at its strongest, the upper tropospheric flow is markedly zonal, the Rossby waves are few and shallow. For us at the surface this means unsettled 'westerly' weather, a rapid succession of lows and highs from the Atlantic, much wind and rain and temperatures above normal in winter and below normal in summer. In this situation there is minimal exchange of air across the lines of latitude and clearly the atmosphere is neglecting its task of heat and momentum transfer.

At the other extreme, the vortex is weak, the Rossby waves exaggerated and there is much meridional (northward or southward) flow. Great highs appear and remain static for long periods, athwart the usual path of the westerly flow. Under these so-called *blocking* conditions, warm air encroaches northwards even to Arctic regions while elsewhere cold air thrusts unusually far south into sub-tropical latitudes. The weather we receive in Britain depends on the location of the blocking high but is always persistent and often unseasonable. Such a high over the eastern North Atlantic may leave us shivering in an Arctic airstream, the common blocking high over Scandinavia brings in even colder Pc air in winter (as in 1947 and 1963) but may give heat waves in summer. From the point of view of the circulation, there is

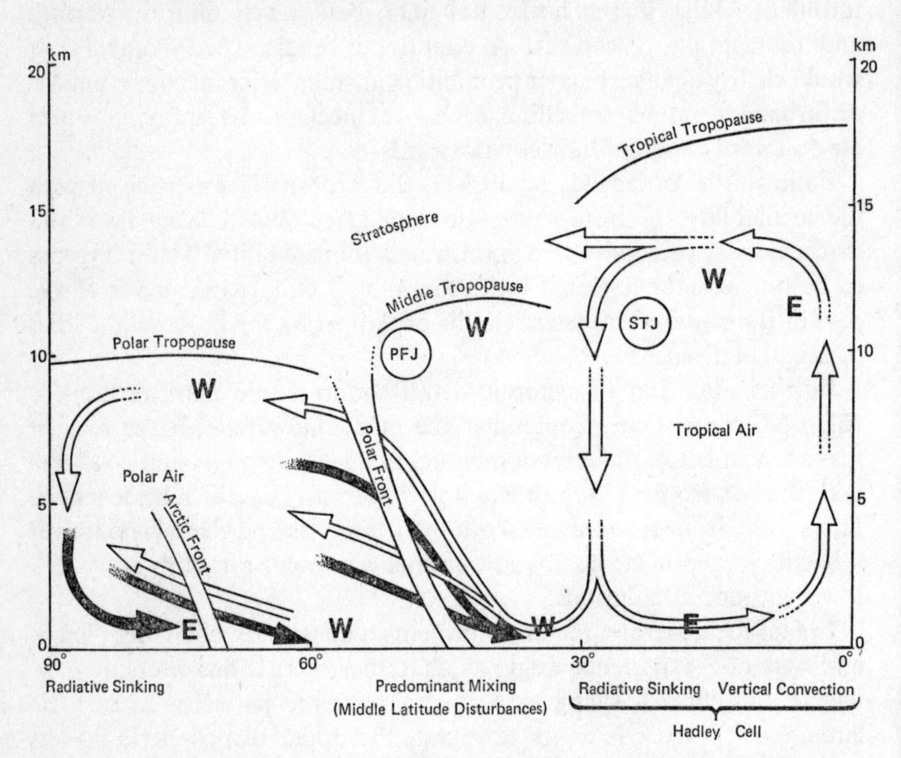

7.16 Some circulation features of the northern hemisphere in winter (after Palmén). Solid arrows indicate cold air.

the maximum cross-latitude exchange of air. It seems that the atmosphere solves its problems spasmodically rather than continuously.

With these complexities and an awareness of still-existing gaps in our understanding, it is not surprising that essays at diagrammatic representation of the general circulation are now notably cautious. Fig. 7.16 is an attempt, based on diagrams by Palmên, to include schematically some of the significant features of the northern hemisphere in winter. It will be seen that a Hadley type cell, which may be regarded as thermally and/or frictionally driven, is retained for the low-latitude circulation, as being necessary here for the poleward movement of energy. The sub-tropical jet stream (STJ) locates near the poleward limit of the upper branch of the Hadley cell, where the northward-moving air has developed maximum westerly momentum through the Coriolis deflection, though Palmén points out also that horizontal temperature gradients are strong in this region. Discussion still goes on as to the validity of other weak or vestigial cells elsewhere but in middle latitudes the predominant impression is one of mixing along gently slanting planes. The apparent penetration of the Polar Front by rising and sinking currents is meant rather to indicate the variability of the front in position and intensity. The Polar Front Jet Stream (PFJ) is found in this zone of strong temperature contrast, strategically placed to evacuate rising air from surface depressions. Other features of the scheme are self-explanatory on the diagram. But it is not to be accepted as applying in all longitudes, or at all seasons, or in any way the last word on the subject.

8 World Weather Types

In the broadest sense the weather of different parts of the world may be understood in terms of their position with regard to the interlocking patterns of air mass incidence, frontal behaviour and circulation types studied in the preceding chapters. Less generally speaking, much will depend also on regional peculiarities on various scales, which may depend on some aspect of the geographical make-up of the region. Weather study properly begins at home and it so happens that, in concentrating hitherto on familiar British weather, we have dealt with a type whose outstanding quality is its changeability. We complain about our occasional cold spells but our worst are not nearly as cold as an ordinary winter in Winnipeg: our hottest spells are cool compared with Bombay in May: our cloudbursts and floods are nothing compared with what may befall the Ganges delta or the Philippines. Our weather in fact is a versatile all-rounder but lacks the high degree of specialization needed for record-breaking.

The more specialized parts of the world are also sometimes the more monotonous and this has often prompted the comment that Britain has weather while many other countries have climates. If the term climate is meant to convey some sort of average or long-term view of atmospheric behaviour, then the implication that in these regions one day's weather is like the average for the month or the season is somewhat exaggerated, though not without a germ of truth. It is the case that some territories lying within air-mass source regions or in steady airstreams like the trade winds have weather that can be described as uniform and reliable compared with ours, but it is equally true that what has come to be regarded as the simplicity of weather in, say, the equatorial and polar regions, is due partly to a lack of knowledge which more recent studies have begun to correct.

In the necessarily brief descriptions that follow it will be convenient to start at the equator, to work towards the increasing complexity of the so-called temperate latitudes and then to return to the comparatively simple regimes of the polar regions. We shall frequently need to refer back to Fig. 7.14.

Equatorial Weather

The basins of the Amazon and Congo, Indonesia, the Philippines, Ceylon, the Malay peninsula and the oceans between about 5°S and 10°N experience weather that we might regard as consistently hot, moist and rainy, although it is not as monotonous as some earlier descriptions suggested. In terms of the general circulation, this is the zone of convergence underlying the ascending limbs of the Hadley cells, the permanent low-pressure belt (equatorial trough) into which trade wind air is constantly drawn from both north and south. In fact, convergence occurs in some areas at some periods, rather than uniformly throughout, and the trough is weak with sluggish and variable winds. At the equator the Coriolis force is zero and even at the margins of the zone the magnitude is often too small to allow geostrophic relationships to develop or circulations to persist.

Air masses in the conventional sense cannot originate here but an all-pervading warmth and high humidity are impressed on any airstream from whatever source that enters and stagnates in this zone, hence the blanket term Equatorial Air (E). On Fig. 7.14 the boundary between the two trade wind streams is shown as a line labelled the Intertropical Convergence Zone (ITCZ) but most authorities emphasize the broad and diffuse nature of this zone and others claim that there are in fact two such convergences, a northern one and a southern one (ITCN and ITCS). Radio-sonde ascents show generally moist and conditionally unstable air, requiring little inducement to generate large clouds and heavy rain.

Equatorial temperatures do not reach the extremes recorded at the tropic of the summer hemisphere but remain remarkably constant, daily temperatures averaging around 27°C (80°F) and varying only a degree or two (C) throughout the year at many places. There is no winter any more than there is a well-defined dry season. The double north-and-south swing of the ITCZ sometimes produces the 'characteristic' equatorial regime of two temperature and rainfall maxima but this effect is slight and easily masked by other factors. In the absence of seasonality, the daily sequence of weather becomes important. The diurnal pressure variation, readily obscured in other parts of the globe, is notable here with differences between the 10 a.m.

and 10 p.m. peaks and the 4 a.m. and 4 p.m. troughs of up to 3 or 4 mb, which may exceed horizontal pressure gradients over large distances.

The diurnal temperature range may be 8° or 10°C (14° or 18°F) and night brings some relief from the sticky heat of the day. With clearing skies the nocturnal cooling may induce saturation in the lower layers of moist air, a heavy dew is often deposited and fog may form, although these conditions are quickly eradicated when the sun rises next day. As the morning wears on, instability soon returns and cumulus clouds appear often building up into heavy cumulonimbus.

The equatorial zone is the wettest in the world, most stations recording more than 1500 mm (about 60 in) and many over 2000 mm (80 in) annually. The intensity of rainfall can be prodigious and Singapore has had 120 mm (4·73 in) in one hour. Such quantities of rain are not to be explained purely by the afternoon convectional storms that have been regarded as so typically equatorial. Thundery outbreaks are certainly common, most of the region having on average 80 or more thunderstorms a year (about 10 times the figure for southern England), mainly occurring during mid-afternoon inland and at night or early morning over the sea. Orographic influences may also be locally important (see Chapter 9) but the heaviest and most persistent rains seem to belong to well organized disturbances which are not necessarily related to the shifting ITCZ, although this does encourage such systems when it lies near the margins of the zone.

Temporary regional or local convergences are responsible for much equatorial rain: these may result from the frictional slowing-down of airstreams at coastlines or from the diurnal local wind circulations that readily spring up where the general pressure gradient is feeble (Chapter 9). For example, convergent sea breezes over islands or peninsulas augment rainfall during the day, while at night the same may happen over water with land breezes from different directions, as is thought to be the case with the violent line-squalls known as 'sumatras' in the Malacca Straits. The interplay of general and local factors makes for a greater variability of rainfall both spatially and in time than is popularly supposed.

Nature responds vigorously to the constant heat and humidity and the natural vegetation cover is the dense luxuriant equatorial evergreen forest. Growth is constant and plants show no seasonal rythms, at least none that are explicable in climatic terms. Insect life is also exuberant and many flies and mosquitoes are disease-carriers. Moulds, fungi and bacteria thrive in what is for them also a highly favourable environment. Man has historically fared less well in the equatorial zone but proper attention to hygiene, sanitation and air-conditioning now make life much more comfortable both for native peoples and acclimatized immigrants.

Tropical Weather

Between the poleward margins of the equatorial zone and roughly 20°N or S lie regions dominated for much of the year by the remarkably steady trade-wind flow but visited in mid-summer by the equatorial trough and the ITCZ. These regions therefore experience more or less of a seasonal pattern since the trade-wind weather of the 'winter' season is mainly fine and dry, while summer brings the sultry heat and rain of the equatorial zone. The tropical summer rain pattern shows some variations as between the islands and windward coasts of these latitudes and the continental areas such as East Africa, parts of Central and South America and Northern Australia.

Over the oceans the trade-wind stream consists mainly of TmK air derived from the sub-tropical highs, though with periodic injections of Pm (see page 173). Warmed from below as they move equatorwards, the moist lower layers become unstable but the resulting convection cloud is limited in vertical extent by a temperature inversion which is one of the major features of trade-wind weather. The trade-wind inversion is due to subsidence associated with the sub-tropical anticyclone and the height of the inversion layer therefore increases equatorwards and also from east (where cold water currents help to intensify the inversion) to west. For much of the region, a blue sky flecked with fair-weather cumulus is typical, with the cloud tops at 2 or 3 km (6 to 10 000 ft), but westwards and equatorwards the tops push further up, with the chances of scattered showers increasing, until conditions merge into the general convectional free-for-all of the equatorial zone.

Local features disturb this pattern sometimes dramatically. Where islands add orographic lift to the unstable air, thick clouds habitually form against windward slopes and these are among the wettest places on earth. In Hawaii an average of over 800 cm (about 320 in) falls on the north-east facing slopes at elevations around 900 m (3000 ft), while the summits of Mauna Loa and Maura Kea (both at about 4200 m or nearly 14 000 ft) receive less than 50 cm (20 in) because they project above the trade-wind inversion, and the south-western slopes are similarly arid because of rain shadow effects. Complex interplay between general and local circulation features may give strengthened winds on windward coasts during the day when sea breezes enhance the prevailing trades while on leeward coasts the two winds may counter each other leaving a relative calm. Lines of cloud often mark the convergences between opposing winds.

The most common synoptic disturbances of the trade-wind belt are the so-called *easterly waves* which appear on the charts as shallow troughs, like extensions of the equatorial trough tonguing into the sub-tropical highs, moving slowly east to west in the deep trades stream. Convergence in these troughs temporarily raises or destroys the trade-wind inversion, giving a period of heavy cloud and thundery rain. A particular significance of these disturbances, already mentioned in Chapter 7, is that a proportion of them grow into hurricanes. The easterly waves have a mainly late summer incidence.

While instability and relief effects may combine to ensure that the trade-wind islands and the coastlands of continental regions have a drier rather than a dry 'winter', the continental regions of the same latitudes have a more strongly-marked seasonal contrast. Warm wet equatorial summers alternate with dry seasons under the influence of land-derived trades or Tc air masses. The length of the dry season diminishes equatorwards but increases polewards until it merges into the constant droughts of the hot deserts under the sub-tropical high pressures, the source regions of Tc air. Under cloudless skies in this dry (though often dust-laden) air much higher temperatures are reached than in the equatorial zone or in the same areas in 'summer': mid-day temperatures exceeding 40°C (104°F) are common, though the same clear skies allow cooling to under 15°C (60°F) at nights. The hottest time of year is immediately before the summer rains and there is often a second temperature peak after the rains cease. The rainy period is hardly more comfortable for being cooler.

Rainfall in these regions is something like 100–150 cm (40–60 in) a year, increasing equatorwards, decreasing polewards. We should regard this amount as distinctly wet in most of England but, in the continental tropics, annual figures for evaporation are of much the same order and the rainfall is less reliable than either in England or the neighbouring equatorial zone: effective drought is thus a very real danger. Whereas the more regularly distributed rainfall of the islands and coastlands supports a natural vegetation of rain forest differing little from the equatorial, the dry season of the continental interiors inhibits real forest growth. Instead we find a sparser covering of stunted trees giving way to the characteristic tall grass of the savannas: in 'winter' these trees are bare of leaf and the grass is dead and yellow.

Tropical Monsoon Weather

The term 'monsoon', from an Arab word meaning season, has come to denote a wind system with a seasonal reversal of prevailing direction: first applied to surface winds, its use has lately spread to the upper troposphere and even the stratosphere, where such reversals regularly occur. The word can describe the wind currents in terms of season (e.g. 'summer monsoon') or of direction (e.g. 'south-west monsoon') or the climatic type or the specific regions in which they occur. The tropical monsoon climate is really only a variant of the tropical summer rain type just described but the summer rain amounts are often extremely large and the transition from dry season to wet may be relatively sudden. The summer rain may be sufficient through its contribution to soil moisture to sustain a dense forest growth throughout the year but where amounts are less a deciduous forest is found. The most notable monsoon lands are in south-east Asia, from India to far beyond the conventional bounds of the tropics in Korea and Japan, but there are important variations within this vast area, as well as many complications.

What is involved basically is an upset of the 'normal' planetary scheme because of a wide extent of land masses in and to poleward of tropical latitudes and the thermal contrasts between land and ocean. It may be simplest to see this first in the case of West Africa. Figure 7.14(a) (January) shows the region covered by mainly north-easterly winds which are clearly trades blowing from the sub-tropical high, hot dry dusty Tc in air-mass terminology but regionally known as the Harmattan. This flow is directed towards the ITCZ which lies at this time very near the Guinea Coast and beyond which is found Equatorial air (E) modified from Tm of the southern Atlantic. In July (Fig. 7.14(b)) the ITCZ has moved far to the north (it can shift even to 24°N), lying along the axis of a thermally induced trough of low pressure (which, it can be seen, extends from a centre over north-west India). The Harmattan has been largely replaced by a 'summer monsoon' flow of warm moist air sometimes described as Equatorial Monsoon (Em) which may be derived from the southern hemisphere trades, turning to become a south-westerly stream after crossing the equator.

An earlier interpretation regarded the discontinuity between Tc and Em air as a true frontal phenomenon, justifying the use of the term ITF, at this remove from the equator, with the cooler Em undercutting the Tc in cold front fashion. However the underlying region remains distinctly rainless – in fact we are deep into the Sahara desert – and it is clear that the air being lifted at this 'front' is too dry to form cloud. Rain is found at this season some hundreds of miles to the south, where the moist Em stream is deep enough to sustain considerable cloud growth and between here and the Guinea coast rain is frequent and often torrential, partly convectional, partly orographic and partly in well-organized disturbances (often generating line-squalls and thunderstorms) in best equatorial fashion.

The more spectacular example of the Asian monsoon attracted much earlier attention, in fact as early as 1686 when Halley first interpreted it as a gigantic sea breeze on a continental scale. This explanation remains widely accepted though in the face of serious objections which suggest that rather more is involved than just the classical thermal mechanism and that a distinctive, if imperfectly understood, rôle is played by the great mountain walls and plateaus of the Asian continent.

The winter pattern has been rather over-simplified as a continental land breeze originating in the thermal (Eurasian or Siberian) high of the interior. It is the case that cold Pc air spreads from this source over northern China giving clear, dry but often frosty weather there, though in crossing the Sea of Japan it loses its inherent stability and brings convectional and orographic rain to the Japanese islands. Part of this Pc stream is brought into confrontation with Tm air at the Polar Front of the western Pacific and disturbances originating here may also affect Japan. Over Malaya and Indo-China, the winter monsoon is composite, being partly 'tropicalized' Pc (bearing no sign of its origin) and partly Tm of the Pacific trades. The pool of dense Pc air over the interior is shallow and clearly cannot cross the formidable mountains and plateaus of Tibet. The so-called 'north-east' monsoon of India, protected by lofty mountain walls on three sides, has therefore nothing to do with the winter anticyclone and is considered to be fed from the sub-tropical high further to the west. Nor is India's 'winter' weather entirely dominated by this stable subsiding air, since shallow lows (possibly of Mediterranean origin) bring occasional but useful rain to the northern plains: somewhat similar depressions appear over south-central China. Ultimately all the different airstreams that constitute the north-east monsoon are directed towards the equatorial trough and the ITCZ, now lying well south of the equator, as Fig. 7.14(a) shows.

The change to the summer pattern begins as early as March in India with a 'spring' (hot-dry or pre-Monsoon) season of increasing heat until in May mean daily temperatures of 35°C (95°F) are regularly achieved. By this time a thermal low (with pressures generally below 1000 mb) has developed over north-west India: as over West Africa, it represents the displacement northwards of the equatorial trough and the ITCZ and the pressure gradient is continuous between this and the sub-tropical high of the southern Indian ocean. No normal Hadley cell is recognizable here. Weatherwise, nothing happens (except for an occasional disturbance or convectional dust storm) until late May or June.

No such dramatic build-up occurs over China and Japan. The polar air of winter gradually retreats, being replaced by Pacific Tm which gives rain when lifted orographically or in the frontal disturbances that form on the northward shifting Polar Front. It will be seen from Figs. 7.14(a) and (b) that Japan and parts of central China see a double passage of the PF during the year, the early summer visitation bringing the so-called 'Plum Rains'. North China and Korea, near the northern extremity of the shift of the PF, receive only a single summer rain maximum, the simplest monsoon regime. The Malay peninsula and Indo-China become covered by a deep layer of hot moist Equatorial air, easily induced to precipitate its moisture by cyclonic disturbances of various kinds, including typhoons (which also affect southern Japan in late summer and autumn).

In India the summer monsoon 'arrives' at the end of May in the south (and in Burma) and advances slowly to reach north-west India only by mid-July. These are average dates: the monsoon is not always punctual. The monsoon (Em) stream may be 5000 m (over 16 000 ft) thick over the Indian Ocean but thins northward to only a tenth of that figure over the north Indian plain. It is highly charged with moisture because of its long passage over low-latitude waters, though not all authorities agree that this air derives from southern hemisphere trades. Much of the rainfall is caused by shallow disturbances known as 'monsoon depressions' which typically form over the Bay of Bengal and then track either westwards or northwards: some develop into intense cyclones. Rainfall amounts depend on factors like the depth of monsoon air and the frequency and tracks of these disturbances: large areas receive only 500–1000 mm (20–40 in) on average and the dry breaks in the monsoon season may spell crop failure, famine and starvation for many. Average rainfall amounts of 200 cm (about 75 in) or more are orographically enhanced, as in the Western Ghats or the mountains of Assam where Cherrapunji has an average of over 1100 cm (about 450 in) but scored double that figure in 1861.

'Autumn' in India (September to December) is the season of the retreating monsoon, when the ITCZ moves southward, generating disturbances that bring important rainfall to the south-east. This is also the main cyclone season in the Bay of Bengal. Thereafter the winter pattern is re-established.

A fuller understanding of this complex sequence of events requires knowledge of upper tropospheric conditions. Reference to Fig. 7.12 shows that whatever happens near the surface, the upper flow is strongly westerly in winter over the whole of Asia: in more detail, contour charts show the sub-tropical jet stream (STJ) divided into two branches, one north of, one south of the Tibetan plateau, much of which extends above the mean elevation of the 500 mb surface. These two branches converge to the east over central China and the associated low-level divergence and subsidence helps to explain the dry winter of this region. In summer (Fig. 7.12 again) the upper westerlies are weaker and have contracted polewards, while an elongated ridge lies between latitudes 20° and 30° with, on its southern flank, easterlies overlying peninsular India and Indo-China. At higher levels (200 mb) this flow strengthens into a definite Easterly Tropical Jet stream (ETJ). This reversal of flow constitutes an upper tropospheric monsoon, which incidentally is out of phase with the surface flow.

The establishment of this easterly flow seems to be a requisite of the summer monsoon pattern and there is some evidence that breaks in the monsoon come with the temporary disappearance of the Tibetan ridge and the associated easterlies and the return of the westerly jets. The Tibetan plateau, at one time considered significant mainly as a mechanical barrier delaying the northward shift of the STJ, is now thought by some to assert itself rather as a high-level heat source. The receipts of solar radiation by this elevated surface create the highest known temperatures anywhere in the world at that (500 mb) altitude. A glance back at Fig. 7.3 will remind us how a high-level thermal anticyclone can form in the upper parts of a warmed air column. The upper tropospheric high over Tibet and the easterlies may be related to this.

Attention has also been drawn to the location of the ETJ which in fact extends over to West Africa. The two major areas of summer rain (not counting those where orographic influences are decisive) could be interpreted as lying below the high-level divergences (hence low-level convergences) to right of the entrance and left of the exit of the jet (Fig. 7.7(b)). The monsoon depressions of the Bay of Bengal, which are not frontal and have no obvious causes in near-surface conditions, are explained by some as due to divergences in the overlying easterlies. In short, monsoons are by no means as simple as they were formerly thought to be and the Asiatic monsoons in particular, despite intensive studies and an impressive assemblage of related facts, retain many elements of mystery.

Desert Weather

It is usual to divide the world's arid lands into the hot or tropical deserts, like the Sahara, the Kalahari desert of south-west Africa, the Atacama of South America and the Australian desert, and the middle-latitude deserts like the Gobi and Turkestan, the high basins of western North America, and Patagonia. The difference is not entirely one of latitude, though the middle-latitude types are sometimes called cold deserts because of their low winter temperatures. The 'tropical' deserts, lying mainly between latitudes 20° and 30°, owe their aridity to the more-or-less complete dominance of the sub-tropical high pressure cells: the mid-latitude deserts, extending polewards sometimes to 50°, are dry because of their inaccessibility to rain-bearing winds, due either to interior location or the rain-shadow effects of high mountain barriers.

Hot deserts like the Sahara, source regions of Tc air, under clear skies and high sun, are hot indeed in mid-summer, when day-time temperatures often exceed 35°C (95°F): the highest officially accepted temperature, 57·8°C (136°F) was recorded at Azizia in Libya. Such temperatures easily give rise to super-adiabatic lapse rates near the ground and cause dry convection and whirling dust clouds while, on a wider scale, they engender thermal lows, but these are shallow and do not break the overall grip of the anticyclonic regime. Even in 'winter' maximum temperatures may reach 30°C (86°F). However, the same clear skies encourage much heat loss at night, when temperatures may be pleasantly cool in summer and distinctly chilly, say only 5°C (41°F), in winter, while frosts are not unknown. Very large diurnal temperature ranges, often exceeding 30°C (54°F), are thus typical of desert weather.

The hot deserts are rarely completely rainless although there are stations in the Atacama where the rain gauges have remained dry for years on end. Most of the Sahara has less than 100 mm (4 in) in an average year, though averages mean little in such a highly erratic regime: this amount is likely to fall during a few short-lived but torrential downpours when moist air penetrates the region from outside. The rare desert rains are dramatic also in their consequences, removing surface material unprotected by vegetation and choking the wadis with muddy torrents. On the northern flanks of the Sahara, a little winter rain will accompany Mediterranean depressions, on the southern flank, some summer rain will be associated with tropical disturbances. The high ground of the central Sahara may catch a little extra rain from either cause at either season.

An interesting variant of the tropical desert is to be found along the western coasts of the continents, in the same latitude belt (but extending equatorward to 5°S in the case of Peru), where meteorological, oceanographical and orographic circumstances combine uniquely to give an exaggerated arid effect. Subsidence in the sub-tropical high and neighbouring trade-wind stream is particularly strong, giving a low and persistent inversion: the prevailing trades are off-shore, blowing the warm surface oceanic water away from the coast, which induces the up-welling of cold water from below (the Peru coastal current): the relief, generally of deep water offshore and narrow coastal plains backed by high mountains or plateaus, fixes, limits and accentuates the phenomenon while effectively excluding outside influences. These conditions juxtapose cool sea and strongly heated land, a situation ready-made for local thermal circulations: frequent sea breezes bring in cool oceanic air whose moisture condenses as fog over the cold current bathing the shore. Inland, the fog quickly dissipates over the strongly-heated land. Thus is produced a strange contrast between wet coastal fog (which in Peru sustains vegetation on the coastal hills) and almost complete aridity only a short distance inland.

The interior deserts of middle latitudes are dry for different reasons at different seasons of the year. In winter the continental highs generate a cold cushion of Pc air with its attendant strong inversion and this effectively inhibits precipitation: occasionally the cold air cushion is dented by an invading frontal disturbance which brings a little snow. In summer, thermal lows may develop but the long land track followed by in-flowing air of whatever origin ensures that it arrives largely bereft of moisture and is soon modified into what is effectively (but not historically) warm dry Tc. Occasional summer rain may fall in thundery outbreaks. The interior deserts may be as hot as the tropical deserts in summer but the winters can be very cold, with extremes influenced by high altitude on the one hand and frost hollow effects on the other (see Chapter 9). Both seasonal and diurnal temperature ranges are large.

The North American deserts are no great distance from the Pacific Ocean, less than 300 km (186 miles) in the case of the Mohave, but high mountains interpose themselves, robbing the moist Pacific air masses of their moisture and holding up the passage of fronts which rain themselves out on the windward flanks. These intermontane deserts span a wide latitude range, from north of 40° to tropical latitudes and winter temperatures vary accordingly with some very low minima in the north, while summer day-time temperatures can be uncomfortably hot everywhere.

Deserts have to be defined in terms of their characteristic vegetation or lack of it. It is not enough to specify a particular rainfall amount since this may be nullified by high values of evaporation and transpiration which depend very largely on temperature. A rainfall of say only 150–200 mm (6–8 in) that allows a scattered scrub if anything in the tropical deserts may be sufficient to support a grassy steppe vegetation in the cooler mid-latitudes. Thus the desert shades through semi-desert into steppe on the cooler poleward margin. On the tropical side the deserts merge into sparse savanna through the gradual increase in (summer) rain. The desert borders in fact exhibit a degenerate form of the neighbouring wetter regime which could be tropical, 'Mediterranean', monsoon or continental.

The largest and most characteristic expanses of steppe are in Central Asia and the interior plains (prairies) of North America. Meteorologically speaking, they share many characteristics of the neighbouring mid-latitude deserts. Winters are dominated by Pc air, to which the American prairies are particularly open, with occasional lows giving snow and blizzards and the prevalent cold interrupted sometimes by warm waves associated with the Chinook (see Chapter 9). In summer, a Tc type of air develops and some rain comes with convectional downpours sometimes with damaging hailstorms: tornadoes also occur at this season in the prairies. Although there is more rain than in the deserts, it is very variable from year to year, a matter of great concern in an agriculturally marginal situation. Given sufficient rainfall and careful husbandry, these are the great cereal lands of the world but on the drier margins or during drought spells, over-cultivation and excessive ploughing expose the top soil cruelly to wind erosion, leading to dust-bowl conditions.

Mediterranean Weather

On the western margins of the continents between latitudes 30° and 40° are regions which in summer lie in the grip of the sub-tropical highs but in winter, as the planetary zones shift equatorward, experience the disturbed 'westerly' pattern of middle latitudes. Summer weather is therefore reminiscent of the tropical deserts in its sun, heat and drought, while winter weather is almost British in its alternation of air masses and passage of Polar Front depressions. These winter rain areas are found confined by high mountain barriers to relatively narrow coastal lowlands in the appropriate latitudes of North America (California), South America (Chile), southwest Australia (around Perth) and South Africa (around Cape Town): our nearest example, the Mediterranean Basin, though regarded as the type-region, is in fact less typical and more complex than any of these.

The Mediterranean is a very popular climate. The average temperatures of the warmest month may exceed 25°C (77°F) in Greece or Spain or interior California, which means that some summer day-time temperatures may soar to 35°C (95°F) and beyond (though it should be remembered that certain coastal stations have relatively cool summers due to cold currents offshore, like San Francisco whose warmest month, delayed till September, has a mean temperature of only 17°C, or 63°F). With pleasantly cool nights, summer weather is reliably fine, hardly spoilt by the occasional shower. Winters, while middle-latitude in synoptic detail, are milder, more like late spring in southern England, and the rain, which generally amounts to something between 400 and 800 mm (say 15 and 30 in) according to location, tends to fall on fewer days than we are accustomed to at home. Small wonder then that wealthy Englishmen have long 'wintered' on the French Riviera, that ever-increasing numbers of Europeans from more sun-starved latitudes join the package tours to the Mediterranean beaches and that Americans flock to California for their retirement.

Synoptically, the weather in the Mediterranean region itself owes much to the unique character of this huge inland sea, biting 4000 km (about 2500 miles) into the heart of the Old World, largely mountain-girt and with a sinuous coastline that intimately relates land and water. On mid-winter mean pressure charts it shows up as a shallow low-pressure area between the Eurasian thermal high and the sub-tropical high over North Africa: this is partly due to the warmth of the sea but attention to detail shows that it is also partly a statistical expression of the frequency of depressions, for this is a frontogenetic zone (the 'Mediterranean Front', see Chapter 6). Air masses may be derived from the Atlantic, Europe or Africa. For much of the time the main winter air mass can only be termed 'Mediterranean air' (Chapter 5), the result of polar air entering and stagnating over the warm sea. Tm or Pmr air may arrive in the warm sectors of Atlantic depressions moving in from the west, but the most common low-pressure systems are in fact lee depressions developed south of the great mountain ranges, particularly over northern Italy where they are called Genoa Lows. When this occurs, cold air is drawn in from western or central Europe through gaps in the mountain wall to form sharply defined cold fronts with the existing 'Med' air. On the other hand, hot dry dusty Tc air from the Sahara will be sucked in to form the warm sectors of east-moving lows of whatever origin.

In summer the North Atlantic sub-tropical high extends its dominance over much of the Mediterranean, spreading stable Tm widely, but, in the east of the basin, pressure is low (related to the Asiatic monsoon low) and a remarkably steady northerly flow brings warm dry air from eastern Europe, sometimes regarded as Tc, over these eastern regions.

In view of this variety of air mass visitations it is not surprising that the Mediterranean is extremely rich in its language of regional wind names. Thus the *mistral* (the magistral or masterful wind) is the powerful cold blast of Pm or A air funnelled through the Rhône valley towards the Genoa Low, the *bora* (north wind) is an equivalent burst of Pc air through gaps further east. These contrast with the Saharan winds, variously called *sirocco* (Algeria), *khamsin* (Egypt) and *harmattan* (Morocco), as well as with the northerly flow of summer (*Meltemi* or *Etesian Winds*) in the eastern Mediterranean. There are many others. Local winds (see Chapter 9) are also of obvious importance in this region of islands and peninsulas, high mountains and deep valleys.

The distinctive regime of the Mediterranean climates evokes distinctive responses in the vegetation and land use. Grass cannot survive without irrigation in summer, to which the traditional Mediterranean response was to move sheep and cattle to higher and sometimes distant pastures for summer grazing, the now-dying practice of transhumance. Bushes and trees with their deeper root systems are better suited and the natural vegetation is a rather scrubby evergreen forest (pines, evergreen oak, cypress) though little of this remains in such a long occupied and cultivated environment.

'Temperate' Weather

The dictionary defines 'temperate' in words such as 'not liable to excess of heat or cold': in this sense, the term can be a misnomer, since the temperate zone, which is usually taken to mean those regions, generally between latitudes 40° and 60°N and 35° and 55°S, with disturbed 'westerly' weather all year round, includes not only fairly equable maritime areas like our own but also vast continental expanses where considerable temperature extremes are encountered. Western Europe, the Pacific slopes of Canada and the States, southernmost Chile, New Zealand and Tasmania exemplify the first category, while the interiors of Eurasia and North America alone represent the second, for the land masses of the southern hemisphere have too little spread in the appropriate latitudes.

The truly temperate climates belong to the western margins of the continents and their weather, changeable, unreliable but lacking violent extremes, is readily understood in terms of alternating air masses and frequent frontal disturbances, as described in the earlier chapters of this book. This kind of weather can only be described synoptically: averaging even over a week may well hide the essential truths that every Englishman knows. The prevailing temperateness stems from the dominance of maritime air masses moving in from the west over ocean waters exceptionally warm for the latitude in winter: the occasional extremes, such as they are, come with invasions of continental air, very cold in winter, hot in summer, or with air unusually fresh from Polar regions.

The continental regions display a fairly simple seasonal pattern, being tightly gripped by very cold, highly stable Pc air in winter developed under the thermal high, while in summer, with pressure relatively low, air masses historically of various outside origins settle and warm up to something akin to Tc. Persistence of weather is one keynote, with cold spells in winter when temperatures remain below zero Fahrenheit (−18°C) for long periods and hot spells in summer, when day-time heat rivals that of the interior deserts. Very large diurnal ranges, encouraged by clear skies, are another feature and the annual ranges are among the biggest recorded anywhere, figures like 28°C (50°F) for Moscow and 37°C (67°F) for Winnipeg being by no means outstanding. Yet another characteristic, which follows from these large ranges, is the dramatically rapid rate of warming up in spring and of cooling down in autumn, in contrast to the gentle transitions of the maritime regions.

Precipitation amounts vary widely in the temperate belt, as does their distribution throughout the year. On the western maritime margins the frequency of unstable air and the procession of frontal disturbances ensure a fairly even spread of rain throughout the year, though usually with the wettest months in autumn and early winter and the driest in the spring. But one does not need to travel very far eastwards in

Europe, even within the British Isles, to find July and August totals not far behind those of October or November: this is the first sign of a continental regime – the ascendancy of summer instability rains – which becomes more and more marked further east. The continental areas acquire a snow cover when energetic depressions from the west penetrate the fortress of the winter anticyclone, but the bulk of the precipitation comes as summer rain in convectional showers, augmented by occasional disturbances. Actual amounts of precipitation vary from over 2500 mm (100 in), sometimes well over, on high ground in western Norway or Scotland or the Lake District to around 500 mm (20 in) in the Thames estuary and lee areas elsewhere: in the continental interiors, something between 400 and 600 mm (say 15–25 in) may be expected. A bigger contrast is in the number of 'rain days' (defined as having 0·2 mm or more): in continental regions, this figure is usually something like 100 per year, while on the western margins it is generally over 150, exceeds 200 in parts of upland Britain and even 300 in certain areas of Chile.

A very different physical grain dictates important contrasts between the Eurasian and North American continents. Atlantic air masses and depressions can penetrate deep into eastern Europe and Asia and, equally, cold winter Pc from the heart of the great land mass can invade as far west as Britain: it is the Mediterranean that enjoys the benefits of shelter by the Alpine mountains. The North American interior is shut off from Pacific maritime influences by the Western Cordilleras but is very open to cold waves of Canadian Pc (which can even reach the Gulf Coast) in winter and also to heat waves in Tm air from the Gulf of Mexico in summer. Any air that is drawn eastwards across the Rockies is modified out of recognition and usually arrives as the unduly warm Chinook (see Chapter 9). Equally, the Pacific coast is protected from Pc in winter.

The east coasts of the two northern hemisphere land masses are somewhat intermediate between the maritime and continental types. The mid-winter circulation map (Fig. 7.14(a)) shows the mean position of the Atlantic Polar Front not far from the eastern seaboard of Canada and the United States: these areas may be cold for their latitudes in winter, being vulnerable to cold Pc from the interior, but Atlantic air is brought in through frontal disturbances, which moderates the temperatures and also gives substantial winter precipitation. Summers are cooler than inland and the PF is still not far away, so that disturbances continue to give rain: these are all-year rainy areas. The coastal regions of eastern Asia are rather different, having more than a touch of the monsoon. Winters are strongly controlled by Pc air from the interior and the Pacific PF is too far away for disturbances to bring in maritime air: in summer the PF and the attendant tussle of polar and tropical maritime air are in full command. The seasons are thus well contrasted and Vladivostok, for example, though a coastal station and at a lower latitude, has a larger annual temperature range than Moscow and similar annual rainfall, though much more concentrated in the summer months.

The natural vegetation of the temperate regions varies (apart from exotic evergreen trees in some exceptionally sheltered western locations in Cornwall or Brittany) from deciduous forests in the west generally, through mixed forests to coniferous trees in less favourable environments and finally to grassland in the drier interiors. In much of the region, the forests have given way to meadows and fields and the grasslands to some of the richest granaries of the world.

Polar and Sub-Polar Weather

The 'temperate' lands of Eurasia and North America are flanked on their poleward side by immense expanses of coniferous forest ('taiga') which range from coast to coast: this region is sometimes known as the sub-Arctic (there is no equivalent in the southern hemisphere). The western maritime fringes are raw, rainy, windy regions frequently visited by depressions born of either the Polar or the Arctic front. The interior regions concede in continentality to nowhere else in the world. The long summer days can still be surprisingly warm, with 20°C (68°F) quite a usual afternoon temperature (as it can be on many an English summer day), but Fairbanks, Alaska (latitude 65°N) has recorded 34°C (93°F) and even the notorious Verkhoyansk (latitude 68°N) is reported to have clocked 37°C (nearly 99°F), not far short of the highest ever recorded in Britain (38°C or 101°F). On the other hand higher latitudes mean longer snow-bound winters and the northern hemisphere 'cold pole' seems to be disputed between two Siberian frost hollow sites, Verkhoyansk and Oimyakon, where the year's lowest temperature (annual absolute extreme minimum) averages below −60°C (−76°F). Verkhoyansk boasts an absolute range (from highest ever maximum to lowest ever minimum) of 106°C (191°F).

Further poleward still (and also at higher elevations within the region just described), as summer temperatures weaken, tree growth is inhibited and the forest yields gradually to the *tundra* lands, snow-covered for much of the year but showing a meagre vegetation of mosses, sedges and stunted shrubs during the brief summer thaw. Beyond the tundras are the permanent snow and ice of Greenland, Baffin Land, Ellesmere Island and the pack ice (not continuous in summer) of the Arctic Ocean. Tundra and ice and snow form the Arctic proper. The southern hemisphere equivalent is the high frozen plateau of Antarctica. Relatively little was known about the two polar regions, except through occasional explorations, until the Russians established their ice-floe stations in the Arctic in the late 1930s and some long-period stations were set up on the Antarctic ice-cap during the International Geophysical Year of 1957.

The two polar regions share similar astronomical conditions but of course, there are important geographical differences. At the poles themselves, the mid-summer day sun never sets and summer insolation is intense. Despite this, the benefits are small, since most of the radiation is simply reflected away unused from ice and snow surfaces and much of the rest is dissipated in melting the superficial layers. So Arctic temperatures in summer are often around freezing point, though may be 10°C (18°F) colder over the Greenland ice-cap but, where the snow melts in the tundra, may reach the levels of a mild winter day in England. In Antarctica, high elevation and albedo keep 'summer' temperatures to −20°C (−4°F) or lower. At the start of the long dark winter temperatures fall rapidly to depths of numbing cold – in the range of −40°C to −50°C (−40°F to −58°F) over Greenland and −50°C to −70°C (−58°F to −94°F) over Antarctica – and stay there until the spring sunrise. The record global extreme minimum, −88°C (−127°F), was recorded in August 1960 at the Soviet Antarctic base Vostok, at an elevation of 3400 m (over 11 000 ft).

It used to be thought that the polar regions were the seats of strong permanent anticyclones but in fact the polar highs turn out to be shallow, weakly developed and very prone to shifting and fragmentation. They are cold-induced thermal highs, feeding shallow easterly winds (the 'Polar Easterlies') at their margins, overlain at no great height by the westerlies of the circumpolar vortex, and can still be thought of as the sources of Arctic and Antarctic air. Some remarkably strong temperature inversions have been measured, often 25°C (45°F) or more increase from the surface to inversion heights varying from a few hundred metres to a kilometre or two. It is also clear that depressions developing on the Polar and Arctic and Antarctic Fronts periodically invade the polar domains and it is the snow they bring that maintains the ice-caps. Precipitation amounts, as far as can be judged from rather inadequate records, vary from 50 to 300 mm (2–12 in) in Antarctica to 100 to 500 mm (4–20 in) in the Arctic. The passage of these depressions or of others nearby can give strong winds that tend to raise clouds of fine dry penetrating snow and this particular combination of cold and wind creates the most lethal environment for unsheltered human beings. The worst blizzards occur when the indraught to these depressions is abetted by katabatic flows (see Chapter 9) at the edges of the Greenland and Antarctic plateaus or funnelling effects between steep slopes.

Mountain Weather

It is hardly possible to speak of a 'mountain climate' in the same way as we refer to a Mediterranean or a Polar climate: what does exist is the modification of a regional climatic regime by what are really local features of elevation, slope and exposure. But mountain weather is a meaningful enough term, as any climber or skier will know. Most of the distinguishing features are due to the rapidly changing form of the land and express themselves as local (or *meso-scale*) circulations of various kinds which will be fully described in the next chapter. Here it would be appropriate to consider some of the effects of sheer elevation as such.

The most obvious is the thinning of the air. At about 2·5 km (8000 ft) atmospheric pressure is three quarters of that at sea level and the partial pressure of oxygen is correspondingly reduced: the tourist leaving his car or motor coach at some of the high Swiss passes, which are not far short of that altitude, may well feel a little breathless and leaden of foot while some visitors are prone to a mild mountain sickness on their way by train to the Jungfraujoch (3454 m or over 11 000 ft) where the average pressure is around 650 mb. Yet there are permanent habitations in the Andes as high as 5500 m (18 000 ft), where the pressure averages only 500 mb, half that at sea level. Indigenous highlanders show a remarkable adaptation to these conditions, having an increased red blood corpuscle count and the ability to breathe deeper and more rapidly: lowlanders can acclimatize gradually to heights of 3000 m (10 000 ft) but have increasing difficulty at higher elevations.

At such heights, especially under anticyclonic inversion conditions, the air is often dry and free of pollutants. Solar radiation is relatively unimpeded and therefore strong and it is pleasantly warm in the direct sunshine of the café terrace, although it may be very cold a short distance away in the shadow of the building. At high altitudes there is more ultra-violet in the solar radiation than lower down and these rays are particularly effective at tanning the skin: strong reflection from snow and ice surfaces enhances this effect. Sunburn is undeniably what many tourists pay for but it can prove painful in excess. A suitable skin cream, as well as efficient sun-glasses to counter glare, should be an essential part of the luggage. The same clear transparent air allows rapid heat loss at night and low minimum temperatures.

Especially in middle-latitude mountains the effect of elevation on temperature may be over-ridden by that of aspect or orientation, so that a high but sunny slope may be warmer than one lower down that remains deep in shadow. Even a cat carefully selects the foot of a sunlit wall for its afternoon nap, so it is not surprising that much of the life and economy of Alpine valleys is adjusted to the pattern of sunny and shady slopes. Not least, the location of the best hotels and holiday villas identifies the most favoured sites. The unequal endowment with radiation at different times of day also results in sharp temperature contrasts which lead to important local circulations (Chapter 9). Aspect becomes less important in the tropics, where the sun is higher in the sky, or in very high latitudes, where diffuse radiation tends to dominate and where the direct summer sun, though low, shines for long periods from the poleward half of the sky.

Mountain weather, at its superb best in anticyclone conditions in dry air, is at other times often cloudy, rainy and windy, since any rain-giving situation is aggravated by orographic lifting. Fronts may be held up for days against the windward flanks of high mountains, while the frustrated tourist watches the driving rain or snow through the hotel window. In middle-latitude mountains precipitation amounts increase generally with elevation but in the tropics there is a maximum at a height of often less than a kilometre above which the upper slopes tend to be relatively dry (page 179). As elevation increases, so too does the proportion of precipitation falling as snow: it is almost 100 per cent at about 3500 m (11 500 ft). The permanent snow-line varies in height from about 5600 m (18 400 ft) above the tropical deserts of the southern hemisphere to sea-level near the Antarctic Circle. It used to be thought that permanent snow beds lay on Ben Nevis but in two recent summers, 1959 and 1969, snow has vanished completely from Scotland: the snow-line (above which accumulation exceeds loss) is estimated to be at 1620 m (5300 ft) in the Ben Nevis area, still some 277 m (over 900 ft) above the mountain top. In unstable summer situations, a combination of convectional and orographic influences can give fierce thunderstorms and spectacular lightning displays with thunder echoing among the peaks are also part of the mountain scene.

9 Local Influences and Local Weather

Air masses, fronts and the controlling pressure patterns are necessary elements in our understanding of the weather of a region but more facts than these may be needed to explain the weather of a particular locality within the region. A variety of local influences, such as the shape and nature of the ground, the proximity of water and the composition of the atmosphere and sometimes several of these combined, contribute toward the realities of weather for a specific place at a specific time. Forecasts issued for the public cover large regions and hardly make allowances for local peculiarities except in a rather vague way. The local factor thus often explains the difference between the forecast we hear or see and the weather we may later experience. Familiarity with local weather therefore enables us to make our own adjustment to the regional forecast. Beyond this, local influences are of specialist interest to people like farmers, gardeners, builders, planners and doctors. Local factors assert themselves most often, though not exclusively, when the general weather influences are weak, which means, in effect, in relatively calm and clear (anticyclonic or col) conditions.

A Matter of Scale

The observant townsman sees that snow or hoarfrost vanishes rapidly from dark asphalt roads but lingers on garden lawns: the farmer knows that some soils are warm and others cold: while picnic parties on hot summer days divide into those that seek the undiluted sunshine and those that prefer the temperature inversion created by the spreading crowns of trees. These are examples of meteorological conditions within small spaces – the air layer within a flower-bed or a forest or the gaps between buildings – where the familiar processes of conduction, convection, evaporation, etc, operate but on a very small scale and in a way closely influenced by factors like the colour, form and thermal properties of the bounding surfaces. This is the realm of *micro-meteorology*, a fascinating subject which we cannot pursue here except perhaps to recall that some of these micro-influences may also affect weather on a larger scale particularly by providing some of the favoured heat sources (see Chapter 4) for convectional activity.

Using a consistent terminology, the general or regional weather is sometimes said to operate on the *macro*-scale, while the local influences with which this chapter is concerned prevail on an intermediate or *meso*-scale. Local variations exist within dimensions of a few to some tens of kilometres: examples are those between coast and inland, hill-top and valley-floor, built-up area and rural surroundings. This scale of study holds special interest to amateur meteorologists because these variations may easily be measured with simple and cheap equipment like the whirling hygrometer (Chapter 1) which, carried on foot or bicycle or car, enables the locality to be traversed in a relatively short time (say not more than an hour) so as to present a near-simultaneous picture.

Coastal Weather

Meteorologically speaking, the surfaces most contrasted in their properties and effects are land and water, for reasons fully discussed in Chapter 3. Over the open sea, where the diurnal temperature change is negligible, there is a complete absence of all those diurnal variations – of relative humidity, of wind and turbulence, of convectional activity – that are so characteristic of inland districts. At sea there are no orographic effects such as there are inland and the frictional resistance of the surface to the wind is very much less. Fundamental differences in weather and climate between land and sea result from these contrasts, as we have seen, but special interest attaches to coastal areas which, being transitional between land and sea, must show in their local weather some of the characteristics of both.

While in detail much depends on the lie of the coast with regard to prevailing winds, coastal districts tend to share to some extent the equable character of marine weather. During the day they are cooler than inland, especially in late spring and early summer (when sea temperatures are rising only slowly compared with day-time inland temperatures) but may be slightly warmer in late autumn and early winter (when the fall in sea temperature lags behind that of the land). At night coastal areas are warmer than inland at all times of year and especially in autumn and winter. Two examples of the effect of the neighbouring sea on coastal temperatures (Fig. 9.1) speak for themselves. It is worth remembering that the North Sea is cooler than the English Channel and our western seas.

Coastal stations have a moderated range of temperature, both diurnal and annual, compared with that inland. They are less likely to have frost and snow. As might be expected, humidities tend to be higher on windward coasts than inland. Winds off the sea blow more strongly and more steadily than further inland and the worst gales are experienced on exposed coasts. In unstable air, windward coasts enjoy no evening clearance of shower clouds. Fog in coastal districts is usually advection fog from the sea (Plate 7) and it may well be slow to shift: radiation fog does not form over the coast but occasionally drifts there from inland. Various orographic effects may contribute to the weather of hilly coasts. High cliffs cause forced ascent and, with moist, stable air, this may result in what is graphically known as *upslope fog* over the high ground: Beachy Head is often hidden in fog of this type. In other conditions, cliffs may squeeze orographic drizzle from stratocumulus clouds or trigger off instability showers or even thunderstorms. Many of these coastal effects may be found on the synoptic charts accompanying Chapter 5.

The most characteristic local weather feature of coastal areas, which exactly highlights their marginal nature, is the diurnal alternation of *land and sea breezes*. These are examples of thermal circulations like those described in Chapter 7, occurring only under quiet general conditions that allow a sufficient temperature contrast to develop between adjacent land and sea strips. As would be expected by reference back to Fig. 7.5, on calm sunny days, when the land develops a thermal low with a complementary high out to sea, the *sea breeze* blows onshore at low levels while there is a compensating flow out to sea above it. At night the circulation is reversed and the low-level flow is offshore (the *land breeze*). Yachtsmen in coastal waters and even inland waters like the Norfolk Broads know these breezes well. Around our shores, such winds are generally too transient to be adjusted to the Coriolis force and are usually more or less onshore or offshore in their direction, but in the more strongly developed circulations of tropical coasts the sea breeze can be deflected to blow parallel to the shore by late afternoon.

The sea breeze is an all-year phenomenon on tropical coasts but in temperate latitudes is generally confined to suitable days in summer. The circulation appears late morning and fades late afternoon or early evening. Recent studies, to which glider pilots have contributed notably,

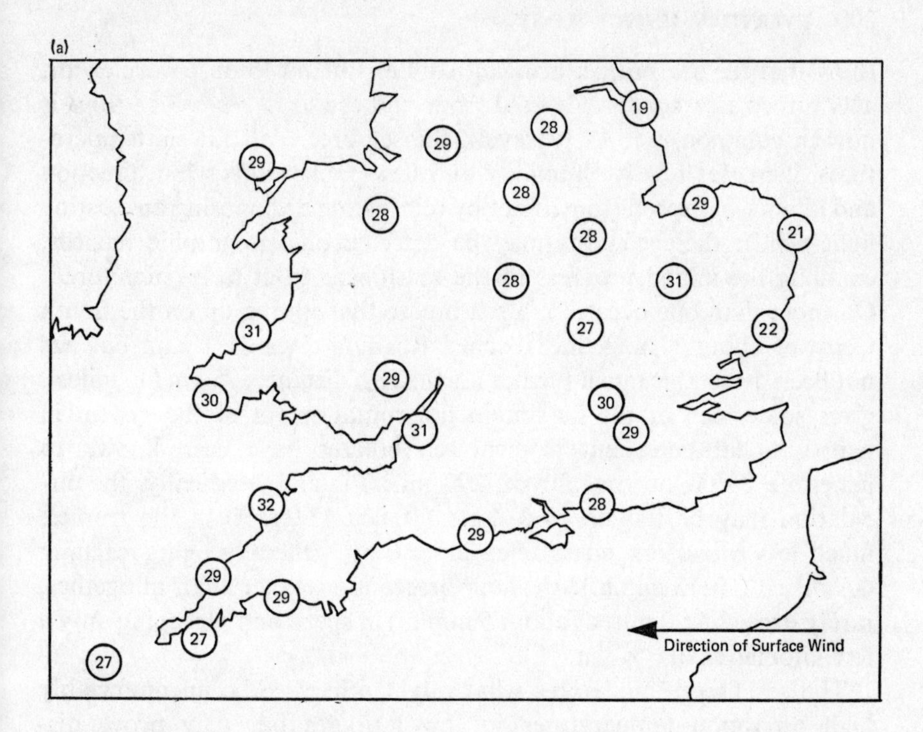

9.1(*a*) *Maximum temperatures (°C) in southern England, 23 August, 1955.*

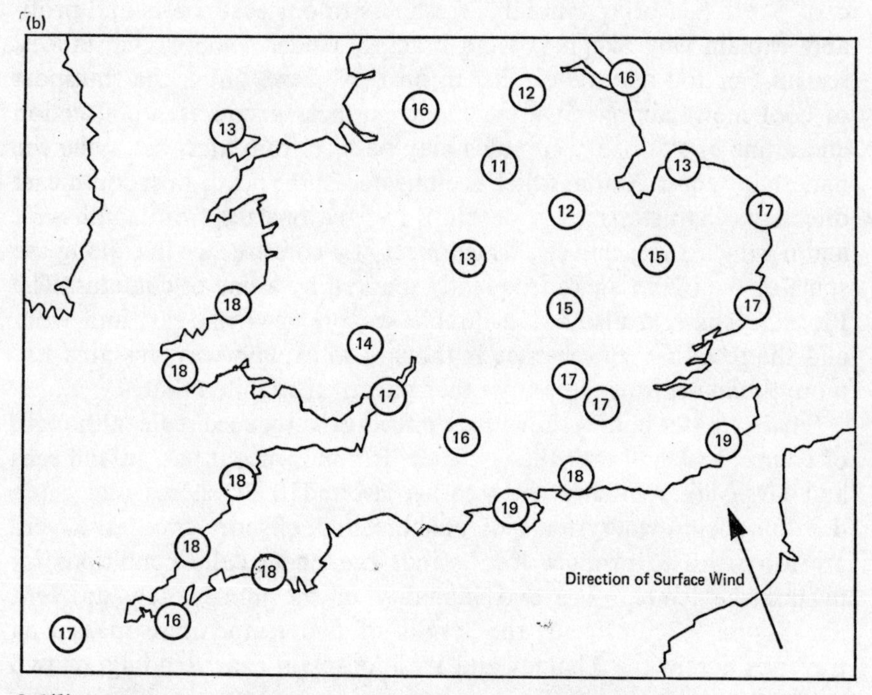

9.1(*b*) *Minimum temperatures (°C) in southern England, 12 August, 1953.*

show that the sea breeze, bringing cool air inland to oust warmer air, acts rather as a small-scale cold front and the term *sea-breeze front* is now in common use. As it travels, the sea-breeze air lowers temperatures, increases relative humidity and dew-point, shifts wind direction and alters wind speed (sometimes by reinforcing or opposing an existing light wind): these changes may be detected on autographic records, enabling the inland progress of the sea-breeze front to be monitored. On more than one occasion, a sea breeze that sprang up on the South Coast at about 11 a.m. has reached Reading towards 9 p.m. but has not been detectable much further north. This distance, 75 km (46 miles), gives some idea of the maximum horizontal extent of the system in temperate latitudes, but tropical sea breezes have been known to penetrate 2–300 km (not far off 200 miles) inland. Vertically, the circulation may be felt up to 3–4 km (10 000–13 000 ft) in the tropics, much less round our coasts. Sea breezes are generally light, reaching 4–7 m/sec (10–15 m.p.h.): the land breeze is a weaker affair altogether, rarely exceeding 2 m/sec (about 5 m.p.h.) in speed and spreading only a few kilometres out to sea.

The sea breeze moderates what might otherwise be uncomfortably high afternoon temperatures in low latitudes but may prove disappointing to our own seaside visitors. Sea breezes off the relatively cool North Sea bring a notable freshness to our east coasts and probably explain why Skegness is so bracing. Under some circumstances, sea mist or fog may be carried in on the breeze. Often the transport of cool moist air over warmed land surfaces encourages convection and a line of stationary cumulus may be seen from afar, betraying the underlying coast. Malta, which is elongated in the north-west-south-east direction, characteristically develops two sea breezes, from south-west and north-east respectively, which meet at a convergence line along the spine of the island again frequently marked by a line of cumulus. The Florida peninsula also has a double sea breeze, from east and west, and the resulting convergence is thought to explain why this area has more thunderstorms than any other part of the United States.

Smaller water bodies show these effects on a reduced scale, although, of course, the shallower the water the less influence it has. Inland seas and large lakes modify the weather around their shores and often develop their own systems of onshore and offshore breezes. Rivers are too small to produce local winds but under calm conditions influence the temperature and humidity of the immediately adjacent air. Figure 9.2 illustrates the results of two temperature measuring traverses across the Thames and its flood-plain near Reading, at two

19. Local weather: valley fog in Edale, Yorkshire.

20. Local weather: shallow ground fog over Stevenage, Hertfordshire.

21. A record snow-drift at Barras railway station, Westmorland, April 1947. The two men are standing on drifted snow 15 or 20 feet above *the railway line.*

22. An avalanche near Davos, Switzerland. The detachment areas can be seen on the upper slopes, left of the picture.

contrasted times of day: it shows what can be done with a whirling hygrometer (and a small boat).

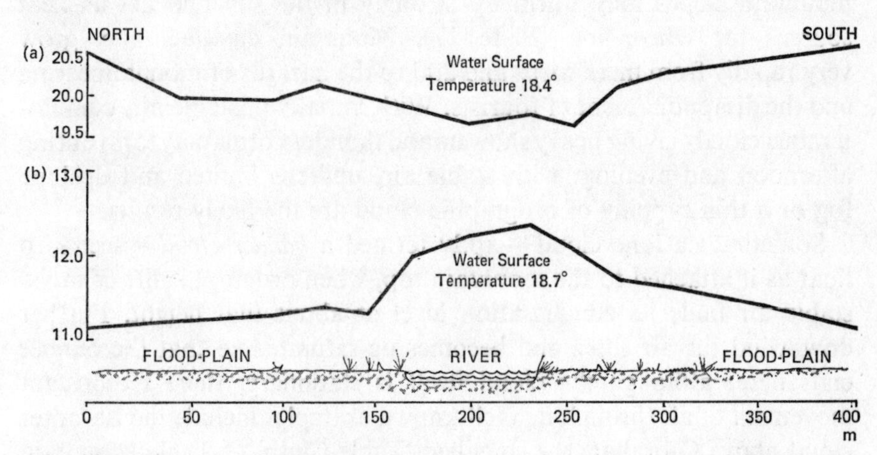

9.2 The effect of a river on air temperatures (a) in the afternoon, (b) late in the evening.

Weather in the Hills and Mountains

The shape of the land creates local weather variations partly by the effect of mechanical interference by high ground on the flow of air and partly through the temperature contrasts that so easily develop in terrain of rapidly changing aspect and elevation. Naturally these local effects are best expressed in real mountain country but some of them are quite well represented among the gentler contours even of southern England.

When air moves over a large obstacle of any kind its flow becomes ruffled by vertical whirls or eddies which are especially persistent on the windward and leeward sides. Such eddies develop, for example, in windy weather, near large structures such as aircraft hangars (thereby increasing the problems of airfield planning) and sometimes (especially embarrassing to architects) in front of tall buildings in new town centre developments, producing unpleasantly strong winds in adjacent pedestrian precincts. They occur too with abrupt natural breaks of slope like escarpments and cliffs and more than one small boy and baffled parent on the beach have released their kite or model aeroplane into the air only to see it crash down time and time again at the foot of the cliff.

The general inducement to forced or orographic ascent in hill and mountain country has already been referred to. Glider pilots make use of these up-currents (*slope soaring*) just as seagulls do above the cliffs. Thermal influences are also at work because, depending on aspect, mountain slopes may warm up strongly in the sun and act as heat sources for convection (Plate 17). Mountain cumulus may grow very rapidly from mere wisps and add to the hazards of mountaineering and the disappointment of tourists. With initially unstable air, cumulonimbus clouds giving heavy showers and thunderstorms may form during afternoon and evening: with stable air, uplift is limited and upslope fog or a thin capping of orographic cloud are the likely results.

Sometimes a lone cloud – aptly termed a *banner cloud* – seems to float as if attached to the mountain top, when orographic lift of moist stable air finds its condensation level at about that height. Further downwind the air sinks and becomes de-saturated so that the banner ends here: although the cloud itself is stationary, there is constant movement of air through it. Well-known examples include the Levanter cloud above Gibraltar, the so-called 'Table-Cloth' of Table Mountain and the photogenic banner of the Matterhorn.

Such orographic disturbance of air flow may be felt at elevations many times that of the high ground causing it. Glider pilots have provided ample evidence of this but the effect becomes visible in the beautiful *lenticular* (lens-shaped) clouds of medium levels (Plate 18). These *hill-wave clouds* are like banner clouds but form in moist stable air layers well above the summits, their gently arched form highly suggestive of their origin. They are represented even at cirrus levels and it is likely that most of the fair-weather cirrus we see is orographic, caused by hills as modest in elevation as the Cotswolds or the Chilterns.

Under similar conditions but with stronger winds, a whole train of standing waves may form in the lee of prominent mountains. These *lee waves* are often clearly betrayed by their accompanying succession of stationary, regularly spaced lenticular clouds (Fig. 9.3). The best known example in this country is produced by Cross Fell which overlooks the Eden valley in a steep escarpment: with a strong north-east wind (the 'Helm Wind'), a hill-wave cloud (the 'Helm Cloud') forms above Cross Fell (but sometimes sits on the summit as a banner cloud) and anything from one to five lee waves, each capped by a cloud (or 'bar') appear over the valley. Satellite pictures elsewhere have shown that lee wave systems can extend some hundreds of kilometres with up to twenty or twenty-five waves. Sometimes, in the immediate lee of mountains, these waves may develop into powerful eddies (*rotors*) with strong up- and downdraughts constituting a hazard to aircraft.

One of the most characteristic mountain winds is the *Föhn* (so called in the Alps) which is usually described as a warm dry wind descending on the lee side of mountain ranges. Such winds can give remarkably rapid and considerable temperature increases (for example, 25°C, or 45°F, in an hour) and can clear snow within a day, hence the Indian name *Chinook* ('snow-eater') in the Rockies. The *Zonda* of

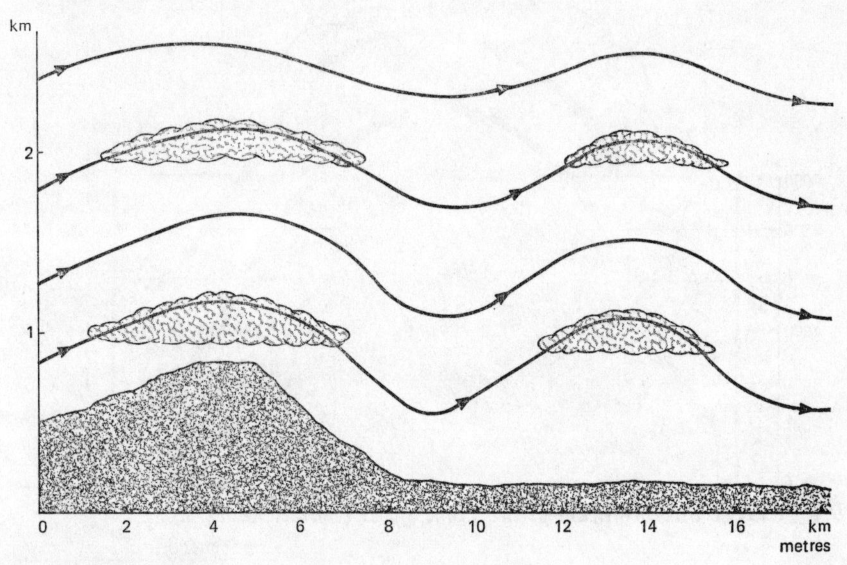

9.3 Wave clouds above and in the lee of mountains.

the Andes and the *Santa Ana* of California are among other regional names for this type of wind. Föhn-like visitations can seriously increase the danger of avalanches, can sadly deplete the ski-slopes on the Cairngorms and necessitate strict notices forbidding smoking out of doors in Swiss villages.

For all its human impact, the Föhn remains meteorologically something of a mystery. The classic explanation of the Föhn, accepted for practically a century, requires a pressure gradient across the mountains, drawing air over so that it deposits most of its moisture as orographic rainfall on the windward side and descends now dry and subjected to adiabatic warming on the leeward side. We can see with the aid of Fig. 9.4, which assumes for convenience a 3000 m mountain, air lifted on the windward side through a condensation level at 1000 m and, following copious precipitation, becoming desaturated at 2500 m on the lee side, that moist air undergoing this treatment cools on ascent largely at the SALR but warms on descent largely at the DALR, i.e.

nearly twice as quickly. In this example the air is now warmer than before by 6°C, having gained the latent heat of condensation of the original moisture. Given the dimensions of our British mountains, even under the most favourable conditions, we can expect an increase of no more than 3° or 4°C through this mechanism.

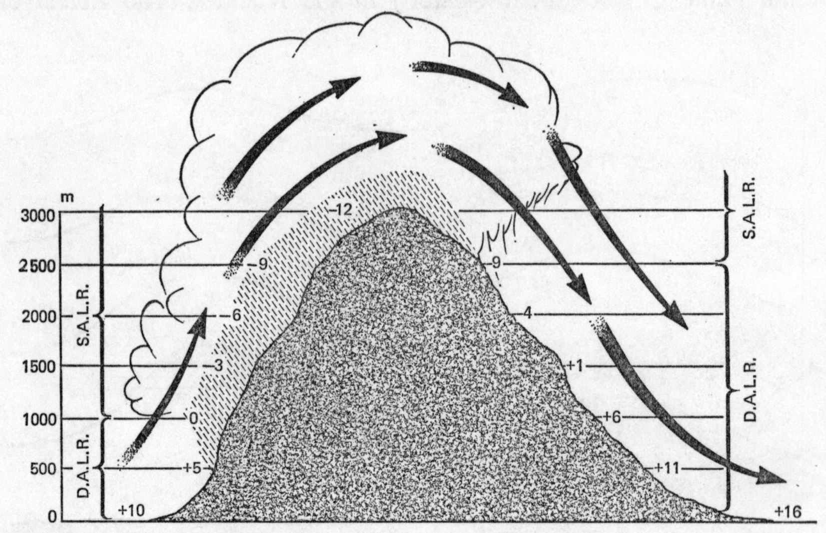

9.4 Classic interpretation of the Föhn effect (temperatures in °C).

There seems no reason to doubt the validity of the classic Föhn concept in some circumstances, as the well-marked cloud edge near the mountain top (*Föhn Wall*) often testifies. But it hardly explains the much more spectacular temperature rises reported in the Alps and elsewhere. For these it is more likely that high level air already warm for its altitude (perhaps in an inversion layer) is brought down and further warmed adiabatically on the lee side of the mountains without being involved in ascent on the windward side at all: it seems necessary anyway to regard the Föhn as part of a forced circulation, a lee eddy, for there is no other reason why warmer air should descend to displace colder. The origin of the air cannot be ignored: thus in the Alps the South Föhn (warm Mediterranean air) is distinguished from the North Föhn (cold European air which even after descent may be only a little warmer than the air it displaces).

Very different from the Föhn and much more like land and sea breezes are the thermal circulations that occur in mountain country due to local heating and cooling. Early in the day the higher sunlit slopes warm up rapidly, warming the adjacent air to temperatures greater than those of the air at the same level high above the neighbouring plains. The ensuing pressure gradients cause ascent of air in contact with the warm slopes and draw in fresh air from the plains to give a general up-valley wind, while a return flow aloft completes the circulation. Within the valley air also rises up the sides and there is a compensating down-current above the valley floor. Large-scale and small-scale effects combine (Fig. 9.5) to produce an upslope wind – the *valley breeze* or *anabatic wind* – which lasts most of the day.

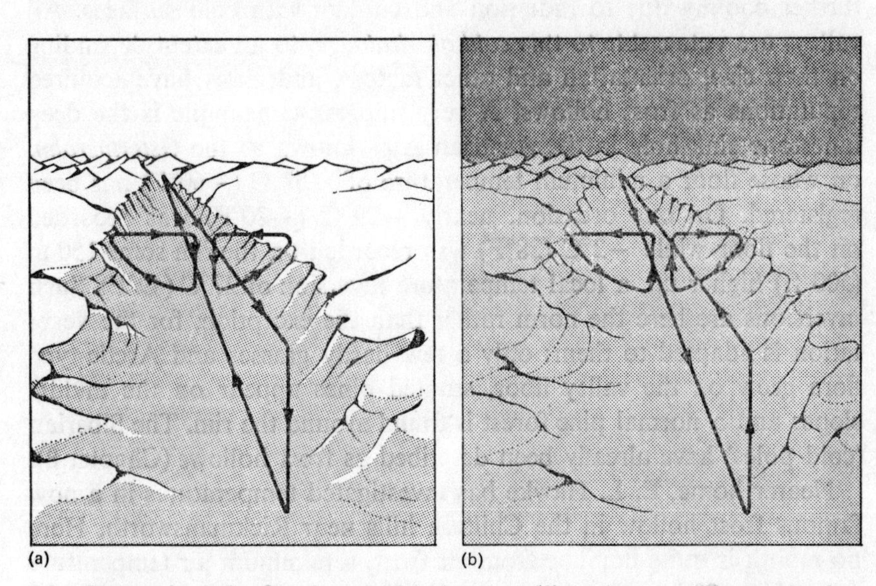

9.5 Mountain and valley breezes (a) day-time (b) night-time. The arrowed lines indicate the longitudinal and lateral components of the circulations.

At night the circulation is reversed. The upper slopes cool first and the air chilled by contact with them drifts down from the valley sides to the floor and generally down-valley to the neighbouring plains. Lateral and longitudinal flows together give a downslope wind which is called a *mountain breeze* or a *katabatic wind*. Stronger in general than the valley breeze and often ruffled into eddies, the nocturnal wind is still light, rarely exceeding 5 m/sec (about 11 m.p.h.). High mountains are not essential for the creation of katabatics, which develop wherever slopes are appreciable. Good examples have been studied on the Chiltern slopes and the amateur can often detect them by the drift of smoke from bonfires or chimneys. A special case of the mountain wind belongs exclusively to Alpine-type country: where snowfields and glaciers cover the higher slopes they provide a constant supply of cold heavy air which descends as a *glacier wind*, reinforcing the normal mountain breeze at night and sliding down beneath the valley breeze during the day. In large Alpine valleys the convergence above the valley floor at night (Fig. 9.5) can lead to cloud formation and even nocturnal thunderstorms: the same has been reported from the edge of the Great Plains as a result of large-scale katabatic flows from the Rockies.

Gravity flows of this kind inevitably accumulate cold air over low ground, any adiabatic warming during descent being far outweighed by further cooling due to radiation and contact with cold surfaces. All valleys are vulnerable to this *cold air drainage*, to an extent depending on their size, orientation and other factors, and many have acquired reputations as frost hollows. A very impressive example is the deep limestone sink-hole in the Austrian Alps known as the *Gstettneralm,* on whose floor a minimum temperature of −51°C (−60°F) has been registered. On one occasion, nearly −29°C (−20°F) was recorded on the floor while −2°C (28°F) was recorded on the rim some 150 m (500 ft) higher up, a local temperature inversion of 27°C (49°F). Such inversions are here the norm rather than the exception, for the vegetation is adapted to them: only a few hardy grasses and Arctic-type flora grow on the valley floor, stunted pines appear on the middle slopes and a normal pine forest is found around the rim. The Siberian 'cold poles' have already been described as frost hollows (Chapter 8).

Nearer home, E. L. Hawke has investigated temperatures in a now famous frost hollow in the Chiltern hills near Rickmansworth. Here no month is quite immune from air frost, a minimum air temperature of −18°C (0°F) and a lowest grass minimum of −21·9°C (−7·5°F) have been recorded. At the same time some high maximum temperatures are on record, for the same configuration that makes a valley a frost pocket on clear nights may also make it a sun-trap on clear days (discounting a valley opening to the north). In Mr. Hawke's valley diurnal temperature ranges have exceeded 22°C (40°F) and even reached 28°C (50°F). In less extreme circumstances, night temperatures at valley floor sites are commonly 3°C, sometimes 6° or 7°C (say up to 12°F) lower than those at hill-top or plateau stations not far away. To farmers or market-gardeners, such variations may mean the difference between suffering or escaping frost damage. Nor must it be forgotten that colder air is nearer to saturation and valleys are therefore especially liable to become enveloped in local mists and fogs which do not affect the higher ground (Plates 19 and 20).

The general relationship between high ground and precipitation amounts was referred to in the previous chapter and the Föhn effect, even if inadequately understood, helps to give meaning to the term 'rain shadow'. So it is that the distribution of rainfall largely echoes the relief of the ground, as a glance at a rainfall map of the British Isles will confirm. With low hills and rolling country, the expected pattern of increased rain on the windward slopes and 'rain shadow' on the leeward may not apply. The lee side may in fact be the wetter because strong winds can drive the rain over the hill crests to be dropped in the calmer air just beyond: this applies even more in the case of snow which, as is well known, piles up in drifts on the lee side.

Townsman's Weather

The town dweller (i.e. most of us in the so-called developed world) also experiences distinctive local weather. It could hardly be otherwise, since in the town artificial surfaces replace field and forest, an agglomeration of buildings obstructs and diverts the wind and atmospheric pollution significantly alters the composition of the atmosphere. A walk through the town reveals a highly variable and always interesting micro-meteorological pattern. We see this in the reluctance of snow to melt on north-facing roofs and the persistence of hoar-frost in the shadow of fences. By turning the corner from a street lying parallel with the wind, we may pass suddenly from the near-gale of an urban Mistral to comparative calm. Sunniness, temperature and humidity vary in street, square and park.

Air pollution as such is dealt with separately in Chapter 10. Its contribution to urban weather is largely via its particulate (smoke) content, which depletes radiation receipts (in some cases by as much as 20 per cent annually and 50 per cent in winter), reduces visibility and encourages fog formation by the vastly increased number of condensation nuclei it generates. The smoke pall does however help to maintain higher temperatures at night by absorbing and returning long-wave radiation, i.e. by enhancing the 'greenhouse effect'. To the extent that town atmospheres can be largely cleansed of their smoke content (and are being, in Britain and many other countries), these influences are tending to wane.

The brick-work, masonry, concrete, asphalt, etc, that constitute typical urban materials behave thermally in the same way as bare rock surfaces. They absorb a good deal of incident radiation (especially if they are dark) and warm up considerably, but, because of the high thermal conductivity of these solid materials, much heat is conducted downwards and, as it were, stored up: during the night, this heat moves upwards and is released, maintaining the surface, and therefore, the air above it, relatively warm. It is largely for this reason, aided by the effect of the smoke pall and the heat loss from buildings and industrial activities, that the town tends to be warmer than the surrounding countryside. This temperature differential, familiarly known as the *urban heat-island* is small (perhaps less than 1°C) when averaged through the year, may be negligible under cloudy, windy conditions, day or night, but quite commonly reaches 6° or 7°C (11° or 12°F) on clear radiation nights. Exceptionally, as reports from some North American cities have shown, the difference may be as much as 10° or 12°C (18° or 22°F): this is likely when a snow cover keeps rural temperatures particularly low. North American heat-islands tend to be largest in winter, suggesting that space-heating makes the major contribution: in Britain and Europe, it is more of a summer phenomenon, pointing to heat storage as the significant factor.

Every built-up area generates more or less of a heat-island, which may be easily measured by traversing with a 'whirler'. London has been intensively studied by Professor T. J. Chandler, using much more sophisticated equipment in a specially adapted vehicle. His traverses, combined with fixed station reports, show that the London heat-island, under the most favourable night conditions, takes the form of a 'plateau', with steep edges (i.e. very sharp horizontal temperature gradients) at the suburban fringe and a fairly flat top (little temperature change) over much of the built-up area except for a minor peak over the most densely built-up City and West End areas. In detail, the heat-island is complex, broken by parks and gardens, a mosaic of heat sources and colder pockets.

The consequences of the urban heat-island are far-reaching. It means for example lower relative humidities in town air: absolute humidities tend to be lower also, since, with rain water rapidly draining off underground, urban surfaces are virtually desert-like. Frost and snow are less frequent in towns than outside: spring comes a little earlier and autumn later, in the sense that the frost-free period (i.e. between last spring and first autumn air frosts) is longer, perhaps by a month or more, giving an unexpected advantage to the urban gardener. The heat source effect may be significant, during the day especially, in encouraging convection: isolated cumulus clouds are sometimes seen sitting motionless above power stations (Fig. 9.6), blast furnaces and factories and there have been reports of large clouds building up over towns while the surrounding country remains cloud-free. Considerable interest (but no unanimity of views) attaches to the question of whether, presumably under marginal conditions, the urban warmth (aided perhaps by the extra condensation and even freezing nuclei present in the air through fuel-burning and industrial processes) can trigger off the thunderstorm mechanism and unleash an 'urban rainstorm'.

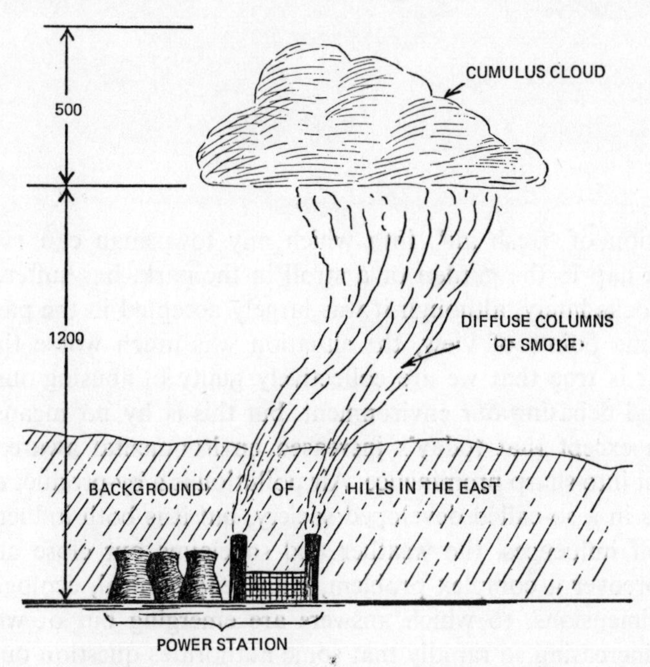

9.6 Cumulus cloud above Brimsdown Power Station (heights in feet).

Some recent studies support this notion, and others suggest that the heat-island can generate a shallow thermal low over the town, with the updraughts of air implicit in the convective mechanisms mentioned above fed by in-blowing surface winds from the rural surroundings.

10 The Abuse of the Atmosphere

The notion of 'fresh air', into which any townsman can escape by taking a nap in the garden or a stroll in the park, has suffered some hard knocks lately, although it was largely accepted in the past when, from some points of view, the situation was much worse than it is today. It is true that we are collectively guilty of abusing our atmosphere and debasing our environment, but this is by no means a new problem except that today's increased environmental awareness has thrown it into sharp prominence. Air pollution is a by-product of man's activities in a so-called developed society, but it is both influenced by, and itself influences, the weather and so claims our close attention. It is moreover a complex problem, with geographical, ecological and social dimensions, to which answers are emerging but of which the scale is increasing so rapidly that some authorities question our ability to apply the remedies in time.

What is Air Pollution?

The term is usually applied to the products of combustion of fossil fuels, whether these are emitted from domestic chimneys, from industrial stacks or from vehicle exhausts. Pollutants are conveniently divided into gases, small light particles that remain suspended in the air for long periods (aerosols) and larger, heavier particles (grit and dust) that fall rapidly and very near their source: the dividing line between these two is generally taken as a diameter size of 10 microns (0·001 cm). Visible pollution (smoke and haze) consists of the larger suspended particles.

When coal is burnt completely, its carbon content is oxidized to carbon dioxide, which is not always considered a pollutant in the ordinary sense since it is a normal constituent of the atmosphere and oceans and enters into plant material (though increasing atmospheric CO_2 concentrations may have implications for climatic change): other main products are water vapour, sulphur dioxide (since coal is about 1·6 per cent sulphur) and the unburnt ash. When coal undergoes incomplete combustion, especially if it is bituminous ('household') coal of high volatile content, the products are carbon monoxide (CO) rather than dioxide, water and sulphur dioxide as before and, more significantly, abundant gaseous hydrocarbons with which are mixed the unburnt carbon particles and tarry globules that make the emission visible as smoke. The ordinary domestic open fire burning bituminous coal in notoriously inefficient fashion is (now that the steam train has vanished) the main, though fortunately decreasing, source of smoke in this country, and in recent years has contributed nearly 10 per cent of the total SO_2 emission (see Table 8). The more efficient factory or electricity power station plant produces hardly any smoke these days but these two classes of coal consumer are together responsible for nearly half the SO_2 output and most of the grit and dust (which require high temperatures and strong draughts for their ejection).

Coke and the other manufactured solid smokeless fuels, by definition produce no smoke, but a little SO_2. Oil is also quite smokeless if completely burnt but, since it has a sulphur content of between 0·75 per cent and 3 per cent, must emit SO_2, the bulk of it from industrial and commercial installations. Hydro-electricity, nuclear electricity and natural gas give no pollution of these forms and are omitted from Table 8.

Another class of pollutants results from motor vehicles. Literally complete combustion of petrol in a motor engine would give only CO_2 and water: the less than complete oxidation actually achieved means that the exhaust emission includes CO in large quantities, various hydrocarbons from unburnt fuel, oxides of nitrogen (mainly nitric, NO) caused by the combination of atmospheric nitrogen with oxygen in the combustion chamber, oxides of sulphur in small amounts and lead compounds (from the 'anti-knock' additives in the fuel) in even smaller amounts. Smoke usually comes only from poorly maintained diesel vehicles. Of the 80 million metric tonnes of oil consumed in 1970–71, only about a quarter was used by road transport but the pollutants are emitted at ground level, the problem is growing fast and there are anxieties concerning the effects on health of some of the emissions.

Partly this is a question of secondary reactions by which the exhaust gases are altered in the town atmosphere in the presence of strong solar radiation. In one of these photochemical reactions, one of the oxygen atoms in nitrogen dioxide (NO_2) detaches itself and combines with an ordinary oxygen molecule (O_2) to form ozone (O_3). More complex products include various aldehydes and nitrates. These are among the constituents of Los Angeles smog which is quite different from the 'classic' London smogs (but more of these later).

Table 8 The Major Polluters, United Kingdom, Year 1970–71. (round figures in million metric tonnes)

Form of Fuel *Class of User*	*Quantity used*	*Pollution Emitted*	
		Smoke	*SO_2*
Coal			
Domestic	20	0·6	0·5
Power Stations	77	small	2·1
Industry	26	0·1	0·7
Coke ovens	25	nil	0·1
Coke and other manufactured smokeless fuels			
Domestic	5	nil	0·1
Industrial	5	nil	0·1
Oil			
All users	80	nil	2·5
	Totals	0·7	6·1

Other Pollutants	
Grit and Dust	0·6 million tonnes
Vehicle Pollutants	
Carbon monoxide	7 million tonnes*
Hydrocarbons	0·4 million tonnes
Oxides of nitrogen	0·3 million tonnes
Oxides of sulphur	0·07 million tonnes
Lead	0·006 million tonnes

*in addition to about 10 million tonnes from domestic and industrial uses.
(Source – *Clean Air Yearbook* 1973)

Awareness of Air Pollution

Coal was in common use in London's furnaces in the thirteenth century and was in fact prohibited by royal proclamation in 1306. Despite this, its use extended to domestic purposes and its visible pollution grew, tolerated by most Londoners but certainly not by all. John Evelyn's famous pamphlet of 1661, *Fumifugium: or the Smoake of London Dissipated* was the first anti-pollution tract, deploring with justifiable passion 'That this Glorious and Ancient City . . . should so wrap her stately head in Clowds of Smoake and Sulphur, so full of Stink and Darknesse . . . ' But the atmosphere of London darkened further and the actuary Frend, writing in 1819, described the 'vomitories of smoke' seen from Blackfriars Bridge, like 'so many volcanoes'. Against general acceptance and indifference, the lone voices were not heard, as other industrial towns followed London's precocious example.

The pollution peak was no doubt reached during the nineteenth century and it was towards mid-century when official anxiety was first expressed in a number of enquiries which led to cautious reports. By the turn of the century, little had been done but at least public opinion was beginning to be informed, helped by the activities of what later became the National Society for Clean Air. Up to 1939, except in relation to noxious gases from industrial processes, legislation had achieved nothing spectacular, beyond permitting local authorities to pass anti-smoke bye-laws if they so wished, but the case against pollution had been powerfully made. Scientific investigation of the problem began, in this country, about 1910 and continued between the wars now largely under Government direction, culminating in a valuable survey conducted in Leicester, of which the now classic report appeared in 1945.

By the mid-twentieth century, it was clear to all who wished to know, that, in its various forms, air pollution is responsible for a substantial loss of winter sunshine in our cities, encourages the formation of fog by adding many more condensation nuclei to the air, is injurious to vegetation by direct attack and by the acidification of the soil through the sulphuric acid brought down in rain, is damaging to limestone buildings (as testified by the semi-permanent scaffolding around the Houses of Parliament) and many materials, is costly in terms of increased cleaning, maintenance and repair bills and wasteful because of the unused energy escaping up the chimneys of coal fires. In the rather different circumstances of Southern California, it was by now established that photochemical pollution causes poor visibility, a characteristic eye irritation (mainly due to formaldehyde and peroxyacetyl nitrate, PAN), damage to crops (due to ozone and PAN) and cracking of rubber (ozone), conditions for which towns like Los Angeles have become notorious.

Not untypically, positive action in this country awaited the highlighting of the problem by a major pollution disaster, the killing London smog of December 1952. No doubt many such episodes had gone unrecorded during the nineteenth century. In December 1930, a 5-day smog occurred in the highly industrialized Meuse valley near Liège, in Belgium: this incident was subjected to close scientific investigation and was shown to be responsible for 63 deaths and many illnesses. In October 1948, a 4-day smog killed 20 people in the metallurgical centre of Donora, Pennsylvania, and nearly half the population of 14 000 were ill. Then came, though not without warning, the London tragedy, with more than 4000 deaths directly attributed to the 5-day smog (Fig. 10.1).

Thereafter, events moved more quickly. A Committee on Air Pollution was set up under Sir Hugh Beaver, which finally reported in 1954. The Beaver Report included a sober but complete indictment of air pollution, incidentally estimating its economic cost to the nation as about £250 million per year (a more recent estimate adds another £100 m to this figure). Out of the report and the increased awareness it helped to create, came the Clean Air Act of 1956 and later the National Survey of Air Pollution.

The Investigation of Air Pollution

The earliest methods of measuring pollution relied on the outdoor exposure of plates or vessels to collect whatever settled by gravitation or was washed down by rain. Of most value for grit and dust deposition, these methods did little for the floating pollutants. A dual-purpose instrument, used to good effect in the Leicester survey and later to become the standard equipment for the National Survey, solved this problem cheaply and fairly reliably. A small pump sucks in outside air via a funnel and tube and draws it through a filter paper held between the jaws of a clamp, so that the particulate content of the air is left as a stain on the filter paper: the air is then bubbled through a bottle containing dilute hydrogen peroxide, with which the SO_2 combines to form sulphuric acid: a gas meter to monitor the air volume drawn through completes the apparatus. The smoke stains are measured

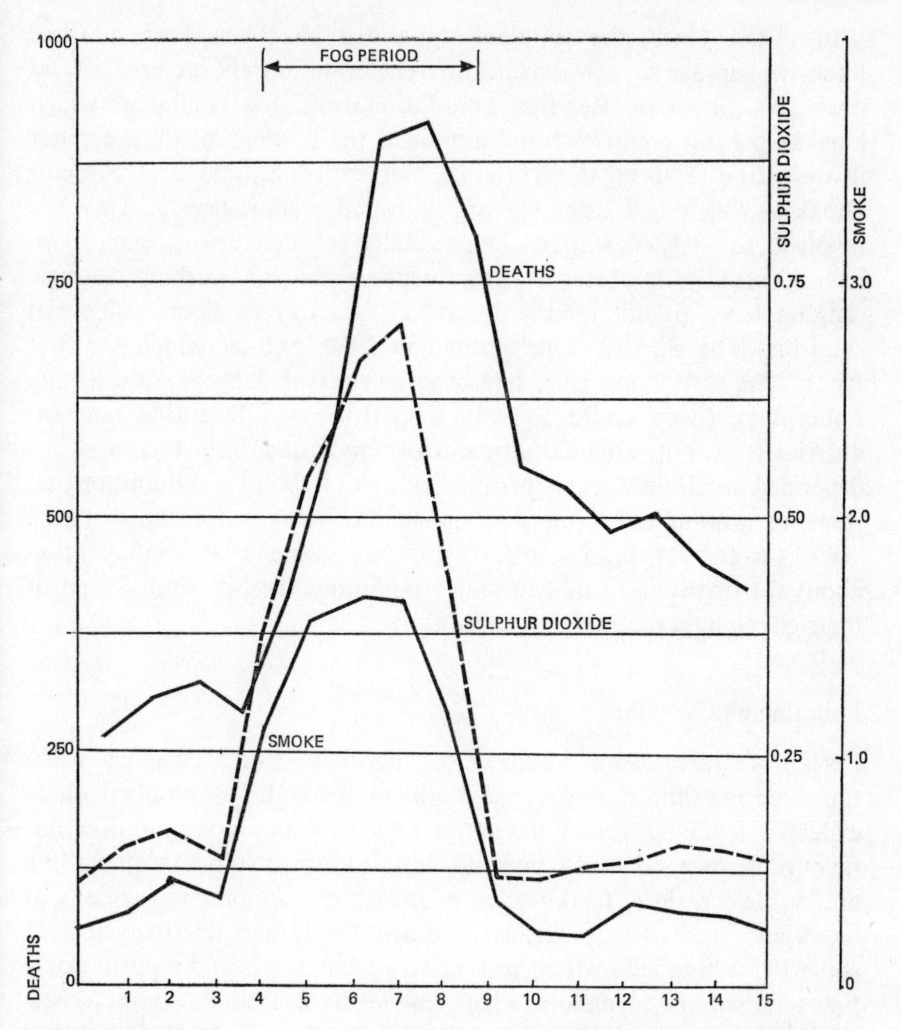

10.1 Pollution concentrations and deaths in London during the first half of December, 1952.

photo-electrically and their density related to smoke concentration in microgrammes/cubic metre (μgm^3). The contents of the bottles are titrated for acidity and the results are translated into SO_2 concentrations expressed in the same units. In the usual smoke filter/volumetric sulphur apparatus, as it is called, the filter paper and bottle are changed daily but there are versions with automatic switching devices which give measurements for periods shorter than a day. More specialist equipment is necessary for measuring the other gaseous pollutants and it is not surprising that much less information about these is available.

In recent years, the standard apparatus has been used in three intensive surveys of towns with different pollution characteristics and problems (Sheffield, Reading and Edinburgh), the results of which have in general confirmed and amplified the findings of the Leicester investigation and in the National Survey of Smoke and Sulphur Dioxide, which has extended our knowledge nation-wide. This has involved some 1200 stations in randomly selected towns, each town having usually five stations representing different urban settings, ranging from predominantly industrial quarters to open residential suburbia. The Survey, which began in 1960, and of which the first reports are now appearing, has been co-ordinated by Warren Spring Laboratory (once under the old Department of Scientific and Industrial Research, now Department of Trade and Industry), and has depended much on the co-operation of local Public Health authorities, some of whom had been keeping similar records for many years. Thanks to this invaluable work we now have some reliable information about the distribution of two major pollutants in this country and of their changing concentration in time.

Pollution and Weather

Certain features were common to the three smog episodes mentioned earlier (and shared by most others), the high pollution emissions expected from an urban industrial area, a valley configuration unfavourable to good ventilation in (the third circumstance) a prevailing anticyclonic regime. In the cases of Liège and Donora, the valleys of the Meuse and Monongahela rivers are deeply and narrowly incised some 100–120 m below their respective plateau levels and a particularly low temperature inversion (which, as we have seen, is characteristic of high pressure) effectively sealed off the valley air. In the much less pronounced topography of the Thames Valley, the anticyclonic grip nevertheless ensured stagnant conditions in a shallow layer, though the top of the fog was over-looked by observers themselves bathed in sunshine on Box Hill at 450 ft (137 m) on the southern rim of the basin.

This underlines the important point that, while pollution is nearly always with us in an urban setting, its virulence is very much a function of the weather. The shape of the land and the character of the town locally play substantial supporting rôles. In broad terms, of course, temperature is the main determinant of the amount of fuel burnt and the annual cycle of pollution shows an obvious winter peak. Over short periods, it is atmospheric behaviour that matters: under what are fortunately normal conditions in this country, pollution is diffused upwards by turbulence and rapidly thinned by the stronger winds at height: but in stable inversion situations, dispersal is inhibited and concentrations mount.

The average daily cycle of pollution in winter shows nicely the interplay of human habits and changing atmospheric conditions: pollution rises to a marked morning peak because fires are being lit or stoked while turbulence has not yet asserted itself, it then subsides to an afternoon trough since turbulence is then at a maximum and fires are low, rises once more to a secondary evening peak as workers return home and fires are revived while turbulence drops, then finally falls to a low night level, when fires are out, despite the relatively still air. The detailed shapes of such graphs vary from town to town and from weekday to Sunday in any one town.

A great deal may be learned about the effect of atmospheric behaviour on pollution by observing the smoke plume rising from an isolated chimney stack: although continuously smoking stacks are not so common these days, if one happens to be at hand, it is worth while for the amateur or school observer to add this to his list of eye observations. In highly turbulent conditions dominated by strong convection, the plume is rapidly spread both upwards and downwards, the large eddies made visible by loops of smoke: this type of dispersion is known as 'looping' (Fig. 10.2(a)). Note that relatively high concentrations of smoke may be brought to the ground quite near the base of the stack. With more moderate eddying characteristic of near-neutral lapse rates, cloudy skies and mainly mechanical turbulence, the plume spreads from the stack orifice in a more regular cone of smaller dimensions, 'touching down' further away and at lesser concentration ('coning', Fig. 10.2(b)).

In inversion conditions (the base of the inversion being above the stack top), there is nothing to disperse the plume vertically and it remains at or near source level, streaming downwind in a thin ribbon, but sometimes fanning out laterally with the vagaries of direction of the light winds ('fanning', Fig. 10.2(c)). If the stack mouth is above the inversion layer (likely only with very high stacks), we have a situation in which dispersal is possible upwards but not downwards: this is 'lofting' (Fig. 10.2(d)). More common and more dangerous is the opposite condition, when the base of the inversion is above the stack top: then upward dispersal is inhibited and the effluent can only diffuse downwards. For obvious reasons, this is called 'fumigation' (Fig. 10.2(e)). It is liable to happen temporarily following an inversion night when the morning sun creates enough warmth and turbulence to stir only the lower air layers. More significantly the 'smog' situations in London and many other cities are a special case of fumigation, in unusually persistent conditions.

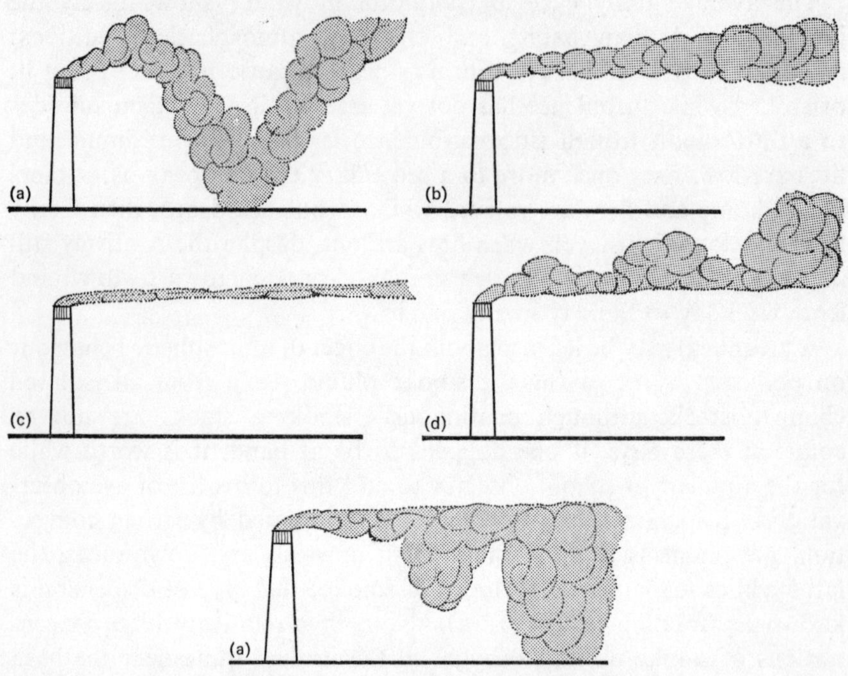

10.2 Observations of smoke plumes from isolated stacks (*a*) *looping* (*b*) *coning* (*c*) *fanning* (*d*) *lofting* (*e*) *fumigation.*

This discussion has assumed an isolated stack and has so far ignored the possible effects of other buildings nearby. These act as obstructions and distort the air flow, particularly producing a so-called 'downdraught' on the lee side. An unwise location of buildings or an unsatisfactory relation of building height to stack height can give local increases of pollution at ground level under certain conditions. The chimney stack will itself create eddies in the air flow, giving 'downwash' on the lee side, which may often be seen and is in any case betrayed by the blackening of the chimney wall. A town is full of pollution sources at various levels, interacting with a complicated three-dimensional structure: it is not surprising that the pollution pattern within a town is highly complex in detail. Taking a broad view, however, it is true that there is most pollution where there are most chimneys, which means that the town atmosphere is dirtiest in its more crowded core of older houses and cleanest in its outer suburbs. This pattern changes little with varying wind direction, not surprising when we recall that dispersion is upwards rather than horizontally down-wind: the town breathes vertically. At the same time, however, there is evidence of pollution drift over long distances, presumably requiring rather special circumstances of persistent air flow under a confining inversion. Traffic pollution is highly localized near its source, falling off sharply within short distances of the main roads.

Conditions in Los Angeles interest us even if only because of the fear that this city's peculiar brand of pollution might occur here. It must be said that very recently secondary pollutants of the Los Angeles variety have been detected here and in view of our fast-increasing motor traffic this seems ominous indeed. However, Los Angeles suffers not only from a super-abundance of automobiles which are certainly the main polluters, but also from persistent inversions (due to the prevalent North Pacific High), the strong sunshine expected in its sub-tropical latitudes (and further enhanced by the anticyclonic regime) which is necessary for the photo-chemical reactions and a coastal basin location that often results in the pollution cloud being carried inland during the day by the sea breeze and back again at night by a combined land breeze and mountain breeze. We are not likely to suffer this particular combination of atmospheric circumstances for more than brief periods.

Towards Clean Air

Through the 1956 Clean Air Act and later provisions, this country may fairly claim to be the pioneer of clean air legislation on a national scale and these measures have contributed to a marked improvement in air quality on a wide though limited front. The Act concerned itself with the visible aspects of air pollution, the reduction of 'dark' smoke (terms like 'dark' and 'black' now being defined by reference to a standard scale of shades). With regard to industry, the Act prohibited the emission of dark smoke (with the exception of certain, very short, permitted periods in unavoidable circumstances), requiring the improvement of firing methods in existing installations and that all new furnaces should be of approved design, capable of virtually smokeless operation. Industry responded well to a call for modernization that was in its own interest as well as that of the community at large, and made the change, often by converting from solid fuel to oil, as required largely by 1961.

The 1968 Clean Air Act extended a provision of the earlier legislation to ensure legal control of chimney heights, empowering the local authority to satisfy itself that new proposals for chimney stacks would not result in downwash or downdraught or other problems of high ground level concentrations of pollutants.

On the domestic front, the 1956 Act allowed local public health authorities to set up Smoke Control Areas, within which the burning of other than smokeless fuels was prohibited. Such programmes, which enabled householders to change to smokeless appliances with the help of conversion grants, were most vitally needed in the more polluted 'Black Areas' but were also vigorously pressed in many other areas. On the other hand some authorities dragged their feet and the 1968 Act strengthened the Minister's hand by empowering him to require local authorities to make Smoke Control Orders. Thus, mainly from about 1960 (bearing in mind that it takes time to implement these orders), the number of Smoke Control Areas steadily multiplied, adding to a few smokeless zones already created by local authority initiative under earlier legislation. Reference to the *Clean Air Yearbook* (published by the National Society for Clean Air) 1973 shows that 29 authorities had completed their programmes by mid-1972 and most of the authorities in the Black Areas had target dates for completion before 1980. The smoke control programme for the entire black areas of England was more than half completed at the end of June 1972: as regards individual regions, Greater London led the field easily (83% of black area acreage under smoke control), followed by Yorkshire and Humberside (60%) and the North Western region (55%), with the rest of the country about one third of the way.

It would be surprising if these developments were not reflected in a marked improvement in smoke emissions and, as Fig. 10.3 shows, the estimated value for the country as a whole has dropped by about two thirds since 1950. It must be recognized however that part of this reduction is due to a general change in the habits of space heating, as people have abandoned the solid fuel fire in favour of the convenience and cleanliness of modern gas or electric fires or central heating by gas, oil, or electricity. Schemes of urban renewal and even minor climatic fluctuations affecting the ventilation of some cities are also possible factors in the general smoke improvement. None of this detracts from the value of the Clean Air Acts in the rôle of pacemakers to what would otherwise have been a slower development.

Turning now to sulphur dioxide (Fig. 10.4), we must not expect a success story quite as striking as that of smoke, since legislation has

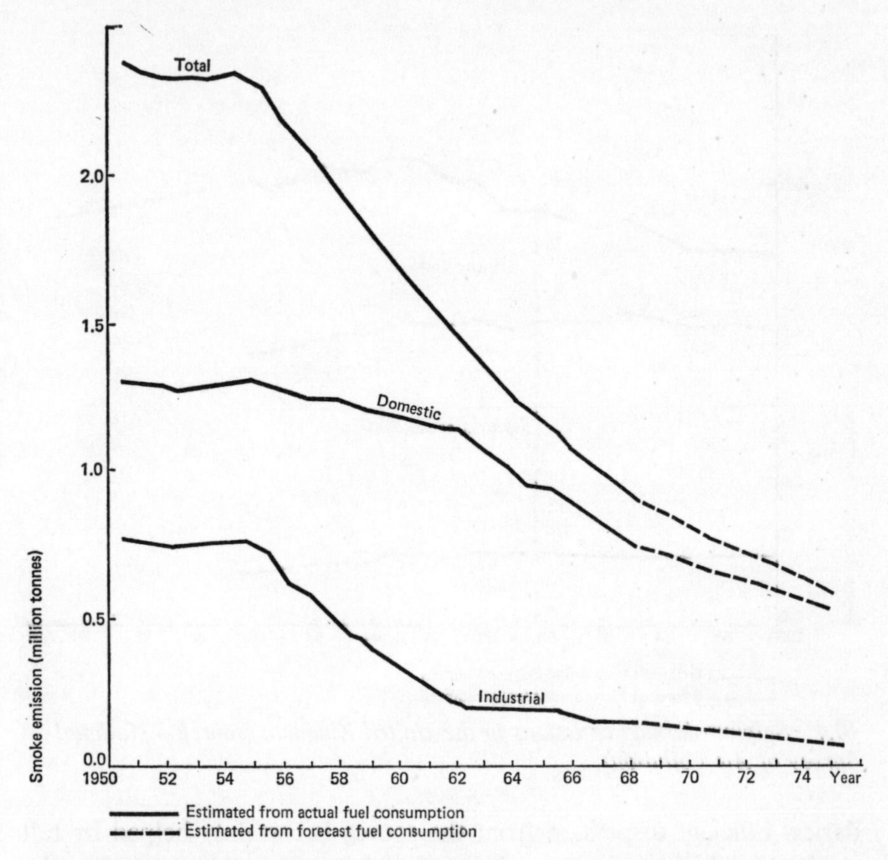

10.3 Smoke emissions in the United Kingdom (source – National Survey of Air Pollution).

not been directly concerned with SO_2 emission, which is inevitable as long as sulphur is contained in the fuel. The slight decline in emission since a peak in 1963 is due to increasing use of natural gas (sulphur free) and nuclear energy. However, the observed ground level concentrations of SO_2 have fallen by about a third in the last decade or so: this reflects a drop in SO_2 emission from low-level domestic chimneys and the effectiveness of the relatively high industrial and power station stacks in aiding dispersion. There is every reason why the trend should continue.

This situation gives rise to some satisfaction but not to complacency. A recent report from Sweden concludes that about half the sulphur deposited on its soil comes from foreign sources (largely Great Britain and Central Europe), through the transport of SO_2 by high level winds and washing out by precipitation. Pollution knows no boun-

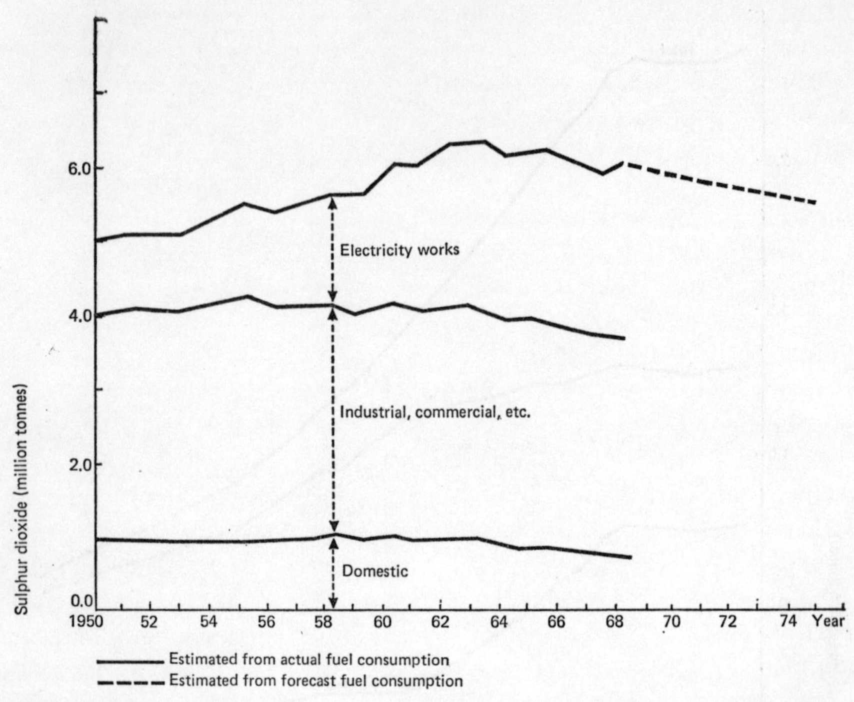

10.4 Sulphur dioxide emissions in the United Kingdom (*source – National Survey of Air Pollution*).

daries. Efficient dispersion from the source is certainly helped by tall stacks and high emission velocity which may double or treble the effective stack height, but appears in part to transfer the problem elsewhere.

We know a good deal about smoke and SO_2 but much less concerning other pollutants, especially those associated with motor vehicles. Without doubt the emission of carbon monoxide has increased steadily in recent decades but little is known of its behaviour in the atmosphere and there are insufficient measurements to establish as yet the rate of increase in concentration. Similarly, data are lacking as regards trends in concentration of the nitrogen oxides.

There is also a large area of uncertainty about what constitutes tolerable limits in the drive towards clean air and the standards at which our efforts should aim. The concept of safe exposure thresholds is well established in factory atmospheres but these refer in the main to relatively fit adults exposed for one third of the 24 hour period in conditions amenable to control and do not necessarily apply to the general community, embracing all age groups and all standards of fitness, continuously exposed to uncontrolled environments. Rarely, in fact, have observed concentrations of pollutants approached anywhere near their threshold limit values, except in the case of CO, of which higher values have been found for very short periods in busy city streets, car parks and, of course, in tunnels.

The effects of CO on the body in various concentrations (acting through the production of carboxy-haemoglobin in the blood) are at least well known. Much ignorance surrounds the health implications of most of the other pollutants. There is currently much anxious discussion on the significance of lead as an exhaust emission, which, although measured concentrations in busy streets fall far short of toxic limits, is known to accumulate in the body. It has not even been possible to pin-point with any confidence the harmful or even fatal ingredients in the London and earlier smogs. The main symptoms were coughing, sore throat, shortness of breath and nausea and the fatalities were among the very old and very young and those particularly vulnerable through chronic illness with asthma or bronchitis. Though little specific can be said, high concentrations of both smoke particles and SO_2, possibly acting together, have been implicated as the main irritant agents. It must be significant that, hopefully the last London smog, which occurred early in December 1962, almost exactly 10 years after the Great Smog, was, in already cleaner air, a relatively minor incident (though it claimed some 750 lives). As regards more insidious and slow-acting influences, there have been enough studies to show strong statistical association between high pollution levels and high incidence of bronchitis, the 'English Disease' of the twentieth century. At the same time, it should not be forgotten that cigarette smoking, itself a highly localized and personalized form of air pollution, stands condemned by a recent report of the Royal College of Physicians as 'now the most important predisposing cause of chronic bronchitis', a major cause of lung cancer and linked also with heart disease, in all resulting in 'premature deaths and disabling diseases' which have reached 'epidemic proportions'.

In the face of all these uncertainties, what now needs to be done in order to give the community the benefit of most possible doubts? Some problems are demonstrably soluble. For example, nearly all the grit and dust from fuel combustion can be removed by the use of suitable arrestors such as electrostatic precipitators (attraction of the particles to one electrode in an electric field) or cyclones (separation of particles by centrifugal force) or various washing techniques. Dust from industrial processes like cement works or thrown up from busy roadways may remain a nuisance locally but at least is not a health hazard (because of the relatively large size of the particles). Any smoke is probably too much but the envisaged extension of the Smoke Control Areas will, through their direct and catalyst effects, reduce the problem to very tolerable proportions and few will grumble at the view expressed by one authority that we should aim at least in the short term at a general improvement to the levels now enjoyed by south-east England (excluding London). SO_2 is a more intractable problem, to which the only final answer is to ensure sulphur-less fuel. Electricity, natural gas, and nuclear power are the bright hopes for the future but oil of low sulphur content is not in plentiful supply. Some sulphur can be removed from coal by washing processes and this is desirable, but methods of washing SO_2 from the chimney gases, such as have been tried at the Bankside and Battersea power stations in London, have not proved rewarding since the washed plume emerges so cold and lacking in buoyancy that it often descends to the ground, the residual sulphur giving higher concentrations than if the gases were left untreated to be released in the ordinary way.

A good clean air principle is that the fewer emission sources there are, the easier they become to control. Thus, in an individual house, central heating (one boiler, one chimney) is an improvement on conventional fires (several chimneys). Similarly, district heating, in which whole neighbourhoods are supplied with hot water or steam from a single boiler-house, is a further advance and there are several such schemes in existence. It is in such contexts that we see close relationships between clean air policies and town and regional planning in general. The careful location of new industry in relation to existing towns, the separation of urban functions within the town (zoning), the preservation of old and the provision of new open spaces such as parks and green belts, insistence on suitable housing densities, the careful planning of chimney sites and heights in relation to buildings following now well understood principles (often based on wind-tunnel tests with models), are among cases in point.

This applies just as much to traffic pollution, which, although localized, is highly concentrated and must appear to the city dweller as growing uncontrollably. In fact, vehicles are already the major source of air pollution in the USA and will become so here. This is a sector of the clean air front in which the USA has taken the lead, at least as regards the restriction of the pollutants at source. Standards of emission of hydrocarbons, carbon monoxide and nitrogen oxides already apply there and certain controls are in force. European countries seem likely to follow this lead, although in Great Britain at present only visible smoke constitutes an offence. The main counter-pollution methods concern engine and carburettor modifications which use a leaner mixture and involve air injection into the exhaust manifold to complete oxidation of the carbon monoxide and hydrocarbons present. Lead-based anti-knock agents have been banned in some Russian cities and their future is under discussion elsewhere: there are feasible alternatives which would slightly increase the price of petrol. A quite different approach is to seek another means of propulsion which is pollution-free: the most likely alternative, the electric car, awaits a development that will ensure an acceptable range.

The planning approach is to keep traffic moving at a steady flow (which actually reduces emissions) and to separate it from other urban activities. Wide, well-ventilated streets, one-way systems, ring roads, over-passes and under-passes aid the first of these aims, traffic-pedestrian segregation, such as in the shopping precincts of new towns and the Radburn type of residential layout, serve the other. These generally constitute good planning practice from other viewpoints. The ultimate solution to the traffic pollution problem may well lie in our acceptance of a radical restriction of vehicle numbers and movement in our cities.

11 Problems of Weather Forecasting

It is far beyond the scope of this chapter to 'teach' weather forecasting. Many months of training are required to produce a professional forecaster and many years of experience to make him a good one. On the other hand, the keen amateur can sometimes achieve limited success at short-period, local forecasting by keeping a weather eye open and a few principles in mind. This may be useful and is good fun if nothing else, and some remarks later in this chapter are designed to encourage it. But, contrary to some often vocal but not very well informed opinion, the amateur with his restricted view of the weather is far less likely to produce consistently good forecasts than the official forecaster with his synoptic view and his increasingly formidable tools of the trade. At another extreme, some people assume that the forecasting problem is one capable of exact solution and that if the forecast goes wrong – as it sometimes does – that the forecaster is necessarily incompetent and the organization plainly inefficient. The main aim of this chapter is therefore to give some idea of the problems facing the forecaster and of what to expect from his forecasts.

As a great meteorologist, Sir Napier Shaw, pointed out some years ago, most scientists who make predictions do so on the outcome of carefully controlled laboratory experiments but the forecaster has to predict the results of vast atmospheric experiments over which he has no control and whose workings he is only now beginning to understand. It is not long since forecasting was entirely empirical, a matter of rough-and-ready rules based on experience. With the impetus of two world wars and the new knowledge gained by advances in observational techniques and superior analytical facilities, some guiding principles are now taking over from rule-of-thumb methods.

It used to be said that forecasting was an art rather than a science, which implied that a good forecaster drew on hidden reserves of experience and 'feel' to make decisions for which the scientific justification was not always obvious. Since about 1960, the subjective element has receded as the computer has come increasingly into its own but the computer output still requires interpretation, the computations do not yet handle all aspects of prediction and are not always fine enough to operate on the meso-scale: in short, the human forecaster is still indispensable. In any case, it is inherently unlikely that weather will ever be mathematically predictable as are, for example, the motions of the stars and planets, for the atmosphere will always have a trick or two up its sleeve, as it were, that cannot be fed into the programme.

It is also worth remembering that forecasting must intrinsically be especially difficult for a region as marginally situated as the British Isles. Stuck between continent and ocean in the restless middle latitudes, we are a battleground for many invading air masses and disputed territory for various pressure systems. Local influences in plenty – orographic, coastal and urban – help to complicate the issues. Perhaps the most confident forecasting claims are made by meteorologists in the wide uniform spaces of the continental interiors. The British forecaster has learned to be cautious.

Understanding the Forecast

A forecast is not a prophecy but a careful statement of the forecaster's estimate of future development. We get the best out of a forecast if we take the trouble to understand it. As another distinguished meteorologist, Sir David Brunt, put it, 'Efficient weather forecasting is in fact a partnership between the official forecaster and the recipient . . . ' It is as well to know that the confidence that the forecaster feels in his own estimate is implied in his choice of words. In a clear-cut situation the forecast is definite: a statement like 'The fine weather will continue for at least another 24 hours' is unequivocal and if it rains next day there is legitimate cause for complaint. But other statements (. . . . 'a chance of scattered showers') indicate a low degree of confidence in an uncertain situation and it is for us to decide whether to take an umbrella or not.

Most terms used in the official forecasts have their everyday significance but a few have a more restricted meaning. 'Cloudy' for example indicates a complete or nearly complete cloud cover but 'dull' is used only when a complete cover is dense enough to give an impression of darkness and gloom even at mid-day. 'Showers' (brief spells of precipitation with clearances between) mean something different from 'occasional rain' (when skies remain cloudy throughout): the first we associate with unstable (often Pm) air, the second with weak fronts or thick stratocumulus cloud. The word 'frost' is not used without qualification and there is a special set of descriptive terms which are explained in Chapter 12. The man-in-the-street will find all he needs to know about the language of forecasts in the booklet *Your Weather Service* (see page 265). For more specialist customers of the weather service there is more of a problem: the forecasts need to be framed in a way that is most helpful to these consumers, who in turn need a certain amount of education to reap the maximum benefits from the service.

It hardly needs stating that forecasts are the more reliable the shorter the time ahead and the more limited the area covered. Short-distance route forecasts for aircraft achieve a high success rate. Local area forecasts are usually on-target or very near but some generalization becomes necessary and the consumer must make allowance for local influences. With regional, let alone country-wide, forecasts it becomes more difficult to satisfy everybody everywhere, even assuming that the broad lines of the forecast are accurate.

In general, short-period forecasts (up to 24 hours ahead) are pretty accurate although there may be still one or two bad ones in a given month. These are due to unexpected developments and must not be confused with situations in which weather events cannot be forecast in any detail: for example, a forecaster may be quite confident that showers will occur over a given area but he cannot predict exactly where and when they will fall and some parts of the area may well escape them completely. Clearly, it is not easy to assess the accuracy of a forecast. Clearly also, we the public are prone to remember spectacular forecasting failures and to ignore the much more frequent successes. This is part of the forecaster's burden. With medium-range forecasting (say up to a week ahead) and more so with the longer-range, it has to be understood not merely that the expectation of accuracy falls off sharply with time but that the emphasis of the forecast shifts from synoptic detail to a broad indication of the general weather type.

Conventional Forecasting Methods

Until comparatively recently the main synoptic tool was the surface chart and forecasting was largely a matter of *extrapolation*, on the assumption that weather systems would continue to move and develop as they had in the recent past, unless there was good reason to think otherwise. With experience and the useful guide-lines provided by the Norwegian air mass/frontal model, a number of general 'rules' were evolved, e.g. that young and mature frontal depressions move in the direction of the warm sector isobars or that older lows slow up after occlusion. Not that all depressions behaved according to these precepts and it was the art of the experienced forecaster to sense the possibilities of new developments.

A large part of the forecasting problem is concerned with the travel of fronts and pressure systems. As a loose generalization, middle-latitude frontal lows are said to move broadly west-to-east. In fact, they have certain favourite tracks, preferring for example to skirt high ground and to be repelled by strong anticyclones. But any particular depression may take almost any track, as Fig. 11.1 illustrates.

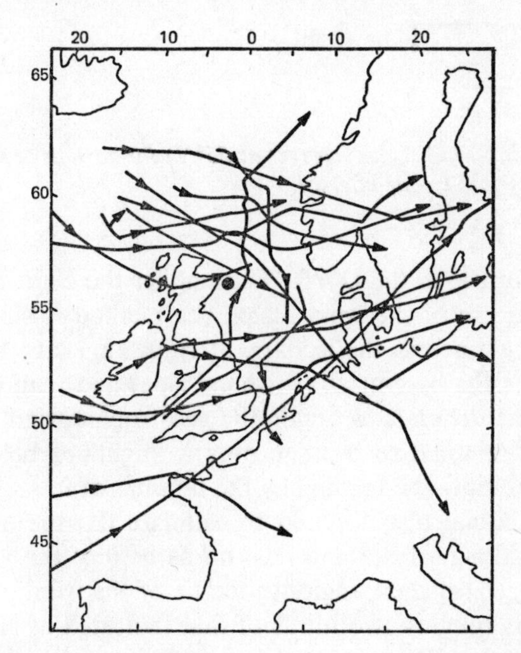

11.1 The tracks of depressions over western Europe during January, 1956.

How can such apparently aimless wanderings be forecast? One well-tried method was to plot the successive positions of the centre at, say, 6- or 12-hourly intervals and extrapolate to fix its future position (assuming no change in speed or direction). This can be used intelligently for curved tracks and increasing or diminishing speeds.

Several techniques, which may still be used by the local forecaster, rest on the fact that the geostrophic wind speed is inversely proportional to the isobar spacing (see Chapter 7). This makes it possible to devise a *geostrophic wind scale,* which, conveniently etched on transparent material, is rarely far from the forecaster's hand. Figure 11.2 shows examples suitable for different scales of chart: the scale of 1:5 000 000 (50 km to 1 cm or approximately 80 miles to the inch), formerly used in the *Daily Weather Report*, is a convenient one for detailed studies of British weather but 1:20 000 000 and 1:30 000 000

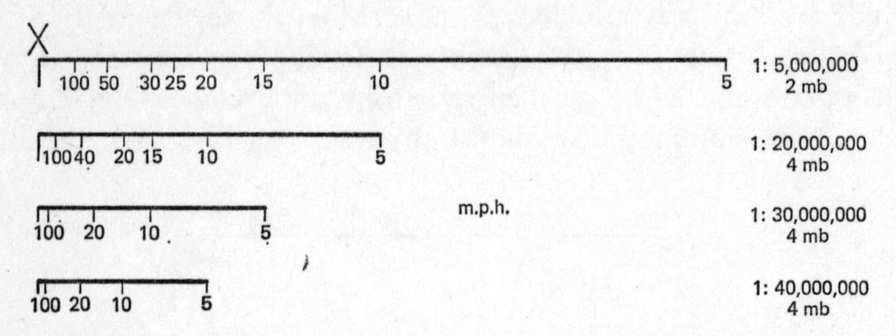

11.2 Geostrophic wind scales (*correct at 55°N*) *for various scales of chart. Note the appropriate isobar interval.*

are currently in use in the DWR. The smaller the chart scale, the less accurate is the use of the wind scale. The method is simple. Lay the appropriate scale across the isobars, at right angles to them, so that one isobar cuts the base-mark (left-hand side) and read off where the next isobar cuts the scale to obtain the geostrophic wind speed in that area. It is more usual to average over a 'flight' of broadly parallel isobars and multiply the reading by the number of inter-isobar spaces. The same scale can also be used to estimate the speed of travel of fronts. Lay it along a front and read off as before: the scale gives the speed of advance (at right angles to the lie of the front). Warm fronts move at speeds roughly two thirds of that indicated by the scale, cold fronts at the scale value.

Use can also be made of pressure changes or *tendencies* (the change, up or down, during the last three hours). Lines of equal tendency – *isallobars* – can be drawn on the chart and show at a glance the present pattern of pressure change: Fig. 11.3 shows the isallobars drawn for the frontal depression of Fig. 6.7. A line drawn from the area of largest positive tendencies (greatest pressure rise) behind the depression to that of largest negative tendencies (greatest pressure fall) ahead of it will often give the direction of movement. Tendencies reflect development as well as movement and, if properly read, yield evidence of the intensification or decay of systems: more than once, falling pressure at a lone weather ship has provided the first clue to fresh cyclogenesis out in the Atlantic.

The introduction of the tephigram (by Napier Shaw in 1925) and of similar diagrams on which aircraft and (later) radio-sonde ascents could be plotted gave the forecaster a much-needed three-dimensional

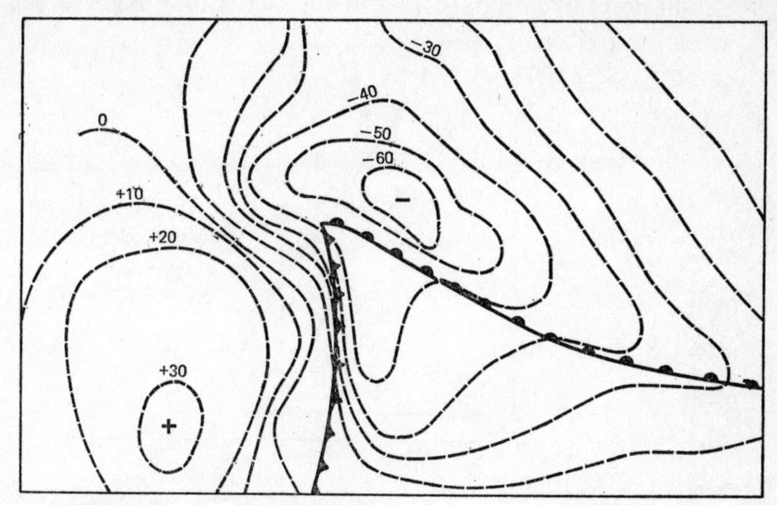

11.3 Isallobars of a frontal depression.

view of the weather situation. It enabled him in particular to forecast the development of weather within the air mass due to movement or, more especially, to diurnal temperature and stability changes. In Chapter 4 we used the temperature–height diagram to depict the development of convection cloud. The forecaster foresees such possibilities long before, usually on the basis of an early morning ascent made under very different conditions. An example will show how this is done.

Figure 11.4 represents a radio-sounding made in south-east England early on a summer morning in 1955. It shows an unmistakable nocturnal surface inversion but, above the stable layer, the air is conditionally unstable for considerable depths and furthermore is very moist throughout (close proximity of environmental temperature and dew-point curves). Here is a situation ominous indeed should day-time warming remove the surface inversion. The surface temperature is bound to rise but how far? With the tephigram (on which area is proportional to energy) there is a technique for predicting maximum temperatures from morning ascents but this cannot be applied to the temperature–height diagram. It would be reasonable to accept a maximum similar to that of the previous day (28°C at Heathrow airport), since the air mass had not changed and a dew-point higher than that at the time of the ascent (because of the evaporation of dew), say 18°C, as it is 30 m above the surface.

With dew-point 18° and dry-bulb temperature 28°, we carry out the construction described on page 73 and Fig. 4.1. Cloud base is reached

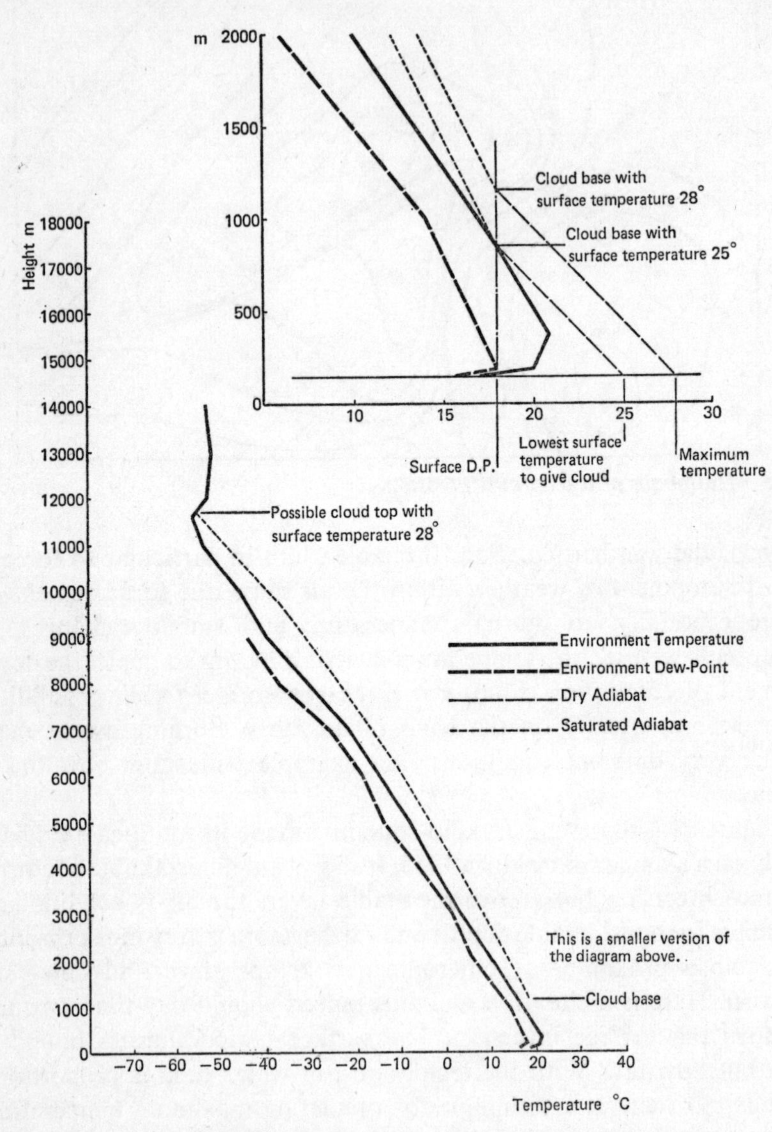

11.4 An example of the use of an aerological diagram in thunderstorm forecasting. Crawley, Sussex, 14 July, 1955, 0212 hr GMT.

at about 1150 m (3800 ft) and the tops would extend to nearly 12 km (say 40 000 ft). Actually, any temperature above about 25°C could set off this powerful convection, in which not only showers but thunderstorms would be more than likely. In fact, 14 July 1955 was notable for severe and widespread thundery activity south-east of a line from the Wash to the Isle of Wight. Six people were killed by lightning that day, three being children struck while sheltering under an oak tree: at the Royal Ascot meeting 44 people were injured (one died later) when lightning struck a metal rail.

Soon after the end of the Second World War, forecasting in this country was advanced by techniques devised by R. C. Sutcliffe at the Meteorological Office, which, using upper air charts as their basic tools, relied on the notion of *development*, which depended in turn on the balance of convergence and divergence at different levels. The thickness charts (page 148) were the key to these methods, from which it was established that surface systems tend to be 'steered' by the thermal winds and that there were certain preferred areas in the thickness patterns for the appearance or deepening of lows or the building up of highs. This involved the concept of *vorticity*, a mathematical expression denoting the spin of the air, in terms of which many of the ideas concerning cyclonic and anticyclonic development described in Chapter 7 can be usefully re-stated. Other notions like the Rossby long waves and methods of predicting their advance also enriched forecasting at this time. Such techniques led to the more reliable formulation of the predicted upper-air and surface charts for 24 hours ahead on which the short-period forecast is based.

Numerical Forecasting and the Computer

Meteorologists have for a long time looked to the objective mathematical prediction of weather as the final answer to their prayers. In 1922 the Quaker mathematician L. F. Richardson published a famous book called *Weather Forecasting by Numerical Process,* in which he stated the problems clearly enough, even if the particular example on which he painstakingly worked (a 6-hour forecast for part of Germany) turned out badly at variance with actuality. This was largely because of inadequate data, but even given a better start, his calculations would not have kept pace with weather events, for Richardson had only a manual calculating machine at his disposal: in fact, he himself estimated that 64 000 mathematicians, similarly equipped, would be necessary to predict the world's weather one pace ahead of its occurrence. Richardson, who died in 1953, as the electronic digital computer was being introduced into weather forecasting, was in advance of his time.

In the computer age, ever-faster machines can perform the calculations in acceptable time and, in fact, from about 1960 in British practice the computer has gradually taken over the key task of preparing the 24-hour prebaratic. Broadly stated, the method predicts the contour pattern at one or more pressure levels for up to 24 hours ahead, through step-by-step calculations of the contour heights at intersection points on a regular grid, using simplified forecasting equations involving barometric tendencies and vorticity considerations, among others. The earlier forecasting models were crude, made big assumptions, for example, that the atmospheric layer in question had no horizontal temperature gradients and that air flow was neither convergent nor divergent, and related to one level only. Restrictive though it seemed, this type of model gave useful results for the 500 mb surface, which, at heights averaging about 5500 m or 18 000 ft, is near the level of non-divergence (i.e. where divergence below gives way to convergence above, or vice versa, in terms of Fig. 7.5).

Later models utilize several levels, feed in more variables to allow for horizontal thermal gradients, the development of systems and the effects of surface friction, and are altogether more realistic. After using a 3-level model with a 300 km grid over an area of the northern hemisphere from Canada to Russia, the British Meteorological Office have lately, with the installation of a giant new computer at Bracknell, changed to a sophisticated 10-level model with a finer mesh. The end-products of the computations are new pressure-fields for the various levels: it remains the task of the forecaster to 'put in the weather' and so to translate into the more familiar features of the surface chart.

It should be appreciated that computers perform not miracles but lengthy arithmetical calculations at very high speeds, obeying instructions fed in by their human masters. Successful forecasting depends on sufficient knowledge of the initial state which is the starting-point of the exercise and on sufficient understanding of the processes that influence the course of events. Forecasts may fail because our understanding is incomplete, because it is difficult to predict new features which are not yet in the initial data (i.e. they are still below the horizon of our awareness), or because of fine-scale systems small enough to slip through the meshes of the grid. Our knowledge is improving (but will never be absolute), observational techniques are advancing (dramatically), computers are getting bigger and faster and the grid is becoming finer-spun. There is optimism that forecasts will improve still further though probably there is least opportunity in the short-range category. But there is no doubt a limit to atmospheric predictability, even if the experts do not always seem agreed as to where it lies.

Further Into the Future

There is an enormous market for reliable forecasts that extend into the medium and long range. Medium-range predictions are of immense value to, say, farmers and builders, who plan many of their activities a week ahead. The gas and electricity undertakings and solid fuel suppliers, river management boards, construction firms, holiday resorts, the manufacturers of ice-cream and cold drinks, among many others, are keenly interested in monthly forecasts and would dearly like to have outlooks for the coming season, if not beyond.

It is feasible to extend both conventional synoptic and modern numerical methods into the medium range. The Americans produce 5-day forecasts by extrapolating the main features of 700 mb contour charts averaged over 5-day periods (but also take into account the behaviour of the long waves and the strength of the zonal flow). British extended-range forecasts, dependent on numerical techniques, cover four days: experimental 7-day forecasts, which also involve the analogue method (see later), are prepared but not yet published.

Although the Americans are experimenting with 30-day mean charts, in general long-term prediction depends on other than synoptic methods. Our Meteorological Office has issued monthly forecasts since the end of 1963, following some years of unpublished trials. These forecasts rest essentially on the belief that meteorological history repeats itself – the so-called *analogue* method. From past records months are sought out with circulation characteristics closely resembling those of the month just ending and the development that followed those months is taken as a guide to the immediate future. Mean monthly surface pressure distributions and 1000–500 mb thickness patterns, as well as synoptic sequences, over the eastern North Atlantic and western Europe, are the analogous features utilized and selection (from something near 100 years of data) is computer-aided.

The method works better for some months than for others: sometimes no really good analogues can be found, at others there may be several which lead to contradictory conclusions. But the objectives of the monthly forecasts are modest, claiming to give only a broad indication of weather type and the expectation of temperature, rainfall and sunshine in relation to average in different parts of the country. The forecasts are revised at mid-month and benefit from the current short-period prognostics. The results have proved more reliable for temperature than for rainfall. As might be expected, the method is rather prone to predict near-average conditions and loath to foresee extremes. The success rate of these very general statements is hard to assess but at least moderate agreement between forecast and fact has been claimed for over 70 per cent of occasions.

Interest is also being shown in *anomalies* in atmospheric behaviour. For example, sea surface temperatures very different from average would be expected to have significant repercussions on pressure fields and related circulation patterns. Some encouraging results have come from relating mean pressure anomalies over western Europe in one month to sea temperature anomalies in the western Atlantic during the previous month. Account is now taken of these temperature anomalies in preparing the monthly forecasts. Recent experimental work has shown the possibilities of predicting seasonal weather (in broad terms) over England and Wales from mean sea-level pressure anomalies over the northern hemisphere during the previous month. For example, very cold springs seem to follow a February pressure anomaly map with large positive anomalies over Greenland and large negative anomalies over southern Europe – indicative of a characteristic blocking pattern (Chapter 7).

There are other approaches, rather more speculative, that seek in past records the clue to future developments. Amateurs have joined enthusiastically in the search for annual recurrences or *singularities,* by which is meant the tendency for weather events to recur at roughly the same time in each year. Little is heard these days of the 'Buchan Spells' (6 'cold periods' and 3 'warm periods', suggested by Alexander Buchan in 1869) but later studies of longer climatic records have confirmed the existence of other singularities. Examples include the stormy sequences that mark rapid changes from anticyclonic to cyclonic regimes, such as often occur during the last week of December (the 'Christmas Thaw') or in late September and March (the so-called 'equinoctial gales') and the tendency to quiet anticyclonic spells in early September ('Indian summer') and just after mid-October ('St. Luke's little summer'). But even though these events undoubtedly occur in a majority of years (though not all), they are too variable in their incidence and intensity to be reliable forecasting aids. The same conclusion must at present be applied to a host of regular *cyclic* fluctuations, with periods ranging from 30 days to many years (including the 11-year sunspot cycle) for which significance has been claimed, though the examination of these periodicities goes on.

The Amateur Forecaster

Amateur forecasting is no doubt as old as human society. People whose occupations keep them out of doors, on the land or at sea, acquire something of an instinct for the immediate weather prospect. Their abilities claim our respect for they have learned by experience to read the relevant sky signs even if they have never heard of warm fronts or unstable air masses and they heed the warning of certain wind directions though they may be quite ignorant of Buys Ballot's Law. At the same time, too much faith must be not placed in them. It is possible at any one time to see perhaps 50 or even 100 miles of sky in any direction, usually no more than a few hours of weather on the way. Sky watching and an intelligent appraisal of winds are useful aids to local short-term forecasting. But the front still far out in the Atlantic, that will bring tomorrow evening's rain, is beyond even the countryman's keen perception though it does not escape the synoptic eye of the professional forecaster.

Both popular experience and popular misconceptions find expression in the proverbs of *weather lore* (of which there is a famous collection by Inwards, while other examples are contained in the 'weather rules' of the Shepherd of Banbury). Some of these weather sayings are well-founded but others have amusement value only. The phase of the moon does not seem to have any useful prognostic value and St. Swithin's Day (15 July) is meteorologically no different from any other: an examination of some London records over a 30-year period showed that the average number of 'rain days' (defined as having at least 0·2 mm) during the 40 days following a St. Swithin's rain day was 14.3 and that following a dry St. Swithin's day was 15·4!

On the other hand, rhymes about red and grey skies have more value, assuming that our weather travels broadly from west to east. Red skies are seen when the sun is low and the atmosphere dry and hazy with many dust particles, as in anticyclonic conditions. 'Red sky at night' suggests that these desirable conditions lie to the west and are on the way, 'red sky at morning' that they have already passed away eastwards. The rainbow, seen from the ground when the sun is low, is sometimes substituted for the red sky in these jingles, none of which is wholly reliable. 'Rain before seven, fine by eleven' works with many cold fronts and weak warm fronts with narrow rain belts.

And when depression follows depression, it is true that 'A nor'wester is not long in debt to a sou'wester'. The interested reader will find it a diverting exercise to examine the weather sayings of his locality and test their validity, as well as to seek their explanation, if any.

Certain animals and plants are sometimes regarded as weather prophets. There is no question that living things may react to current atmospheric conditions – we do ourselves, perhaps more than we realize – but there is little evidence that their behaviour is influenced by weather yet to come. The scarlet pimpernel closes its petals when the relative humidity reaches 80 per cent and seaweed – reputedly the forecaster's standby – dampens at high humidities. Certain chemicals change colour with changing humidity, hence their use in some cheap toy 'weather predictors'. None of this is necessarily prognostic, except in so far as it is indicative of the prevailing air mass.

Ignoring such doubtful aids, what are the prospects of success for the amateur forecaster? They vary of course from situation to situation, but some knowledge of the broad weather pattern and of local conditions provides a good beginning. If we know that we are in Pm air we are not deceived by the clear skies of early morning. The convection clouds will build up sooner or later and the earlier this happens the cloudier the day is likely to be. Since we have no upper-air data we cannot say how much they will grow: we can only wait and watch their vertical development. If they remain flattened (or flatten after an earlier build-up) then there is an inversion aloft: this may be anticyclonic (Plate 8), when a fine or fair day is indicated, or there may be an advancing warm front (in which case look for the tell-tale cirrus wisps). But if the cumulus show considerable vertical growth (less is necessary in winter than in summer) and especially if the tops become frayed (Chapter 4), showers are likely (Plates 3 and 4).

When clouds appear like rocks and towers,
The earth's refreshed by frequent showers.

If we know that Tm or Pmr air is in possession, a dull early morning need give no cause for pessimism. It generally means no more than a thin layer of stratus or stratocumulus and, in spring, summer and autumn, it is likely that breaks will soon appear (the proverbial patch of blue sky 'big enough to make a sailor's pair of trousers'). This will be followed by a more or less complete dispersal of cloud and a warm day. If there is other cloud above, the lower layer is less likely to break up. Another bad beginning that actually portends well is early morning mist or fog for (inland) this often means anticyclonic conditions and (except possibly in winter) should clear during the morning to give a fine day.

If a shallow summer depression spreads from the south, or in the stagnant air of a col, we must bear in mind the possibility of heat thunderstorms. A useful warning is often provided by the cloud type *altocumulus castellanus,* a medium cloud with marked vertical (castle-like) growth, which indicates instability at those levels (Plate 12). With more day-time warming, it only needs convection currents from the surface to penetrate to this unstable layer for convectional overturning to be powerfully reinforced through considerable depths. Many heat waves break up in this way. One investigation showed that, on seven out of ten occasions when this cloud was observed in the London area, thunder occurred somewhere in south-east England on the same or the following day. With large cumulus and cumulonimbus about, it is often rash to predict the usual evening dispersal of convection cloud for these have the habit of persisting and thunderstorms often break out anew in the evening. The reason for this is not clear. It may be that in the evening the cloud tops cool by radiation while the lower parts remain warm so that the lapse rate within the cloud increases and it becomes less stable than before. Or it may be that katabatic downflow from nearby hills causes local convergences and air ascent over valleys and plains.

The most easily identifiable weather system is the warm front because of its fairly characteristic cloud sequence. Fast-moving cirrus is a likely (but not an invariable) herald. After a few hours this should give way to the cirrostratus which is reliably labelled by the halo, often a useful prognostic sign ('The moon with a circle brings water in her beak') although it can of course occur in other circumstances. An early investigation in the London area suggested that there was about a 70 per cent chance of rain within 24 hours of the observation of a halo but later studies do not support such a high probability.

Limitations of Amateur Forecasting

It would not take very long to be convinced of the many pitfalls in the path of the individual forecaster. Even if supposed warm front cirrus merges into cirrostratus which then thickens into altostratus according to plan, there is no guarantee that rain will follow. The front may be weak and there is no knowing what kind of air lies behind it. A cold front may be followed by a marked wind veer, rapidly rising pressure and cold showery Pm air from a north-westerly quarter: or, as sometimes happens, by only a slight wind change, little rise of pressure (even a further fall) and mild Pmr from the south-west. In the latter case there is often a large trough of low pressure yet to come, which may well bring more precipitation than the front itself. We cannot know in advance if a frontal low is occluded or not. In fact there are many situations in which our unaided efforts are quite insufficient: we need to know the synoptic pattern.

It helps to possess a barometer or better still a barograph. With pressure the changes (barometric tendencies) matter, not the absolute readings. If recognition of the warm front cloud sequence is reinforced by a falling barometer and a backing wind, we may be more confident in our prediction of rain. The quicker the pressure change, the more vigorous – and short-lived – is the weather change.

Long foretold, long last,
Short notice, soon past

is often applicable to pressure changes. Persistent systems like warm anticyclones announce their coming by slow changes of pressure. A falling barometer and a freshening east wind generally spell a good deal of rain ('Rain from the east, twelve hours at least'). An appeal to Buys Ballot's Law tells us that pressure is low somewhere to south or south-west of us and quite probably there is most of a frontal depression yet to come. By application of the same law, we can see where we are in any pressure distribution.

The wise amateur acknowledges his limitations and takes advantage of all available information. *The Daily Weather Report* (Chapter 2) gives the immediate history of the present situation and shows broadly what to expect. The forecast charts for noon published in some morning papers give the best opinion available at the time of issue and of course both these 'prebaratics' and actual synoptic charts are shown regularly on BBC television. However, as things stand at present, the early evening forecast is too early for many viewers and in any case is allowed so short a time that it is possible to miss the salient points through a distraction as fleeting as a sneeze, while the late and more leisurely forecast may well be too late for some practical purposes.

In default of such information, the weather bulletins for shipping on Radio 2 (1500 m only) now at 1.55 and 5.55 p.m. (and two other less convenient times) make it possible to build up a rough and ready synoptic chart. The broadcasts give the general synoptic situation of a few hours earlier, including the location and expected movement of highs and lows, and reports of wind direction and force (Beaufort), visibility, weather and pressure at 12 selected coastal stations as well as forecasts for the standard shipping areas. By dint of hasty plotting, using the common symbols, it is not difficult to construct a chart like that of Fig. 11.5, which gives a picture for roughly the same time as

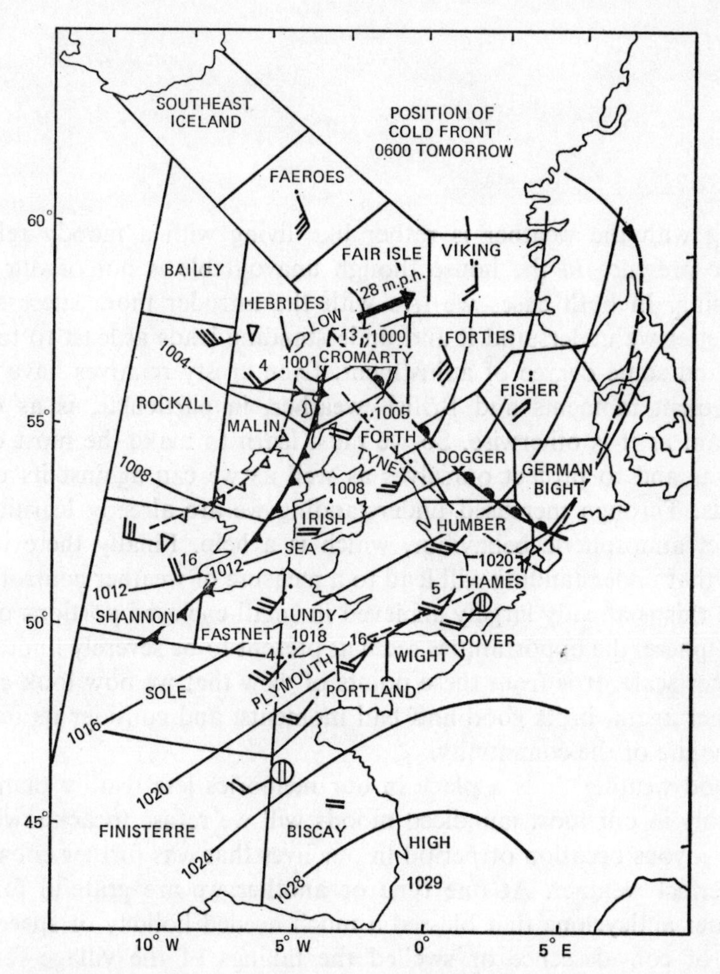

11.5 Synoptic chart for around mid-day, 9 May, 1956, constructed from the early afternoon shipping forecast.

the newspaper prebaratic and can be used to check it. An elaboration of this do-it-yourself weather map was described by C. E. Wallington in various issues of *Weather* (see page 265), using a prepared outline chart, and these articles combined in booklet form as well as the charts themselves are available from the office of that journal.

12 Living With the Weather

Living with the weather is rather like living with a moody relative whose presence in the house though unavoidable is not of our own choosing. In both cases we live with the intruder more successfully the better we understand it and understanding leads at least to toleration and some degree of adjustment. Even crusty relatives have their benevolent moments and British weather, in particular, is as often pleasant as it is otherwise. So we must learn to make the most of its benefits and to protect ourselves as well as we can against its uglier moods. Through increased understanding we are already learning to predict atmospheric behaviour, which is a help. Finally there is the hope that understanding will lead to a measure of weather control and, while this is already largely achieved in small enclosed portions of the atmosphere, the opportunities seem at present to be severely limited on a larger scale. It is from these points of view that we now look at the weather again, in its good and bad moments, and consider its impact on the life of the community.

Good weather finds a place in our memories less readily than bad yet only in our most jaundiced moods will we refuse to acknowledge some joyous occasion or period in our lives that was further enhanced by perfect weather. At one time or another we are grateful for the summer anticyclone that blessed a much needed holiday or speeded a spell of convalescence or swelled the takings of the village fête or Sunday outing. In good weather, other things being equal, life tends to go according to plan.

Townsmen who are not also gardeners sometimes forget that what is good weather for them may be most unwelcome in the countryside. Farmers need dry weather in spring for sowing and in mid-summer for hay-making: they hope for sunny spells for ripening grain and gathering in the harvest: they wait for quiet weather for spraying with insecticides and weedkiller and pray for a rainless period to follow. But equally, rain in moderation is necessary for germinating and sprouting in spring and later for the growth of lush grass and the achievement of good crop yields. Rain, of the right type and amount and at the right time, is part of good weather on the farm.

It is bad weather that makes news. It is a point worth debating whether, with our advanced technology, we are more or less vulnerable to meteorological catastrophes today than in times past. Certainly no year passes without the weather exacting its toll in human lives and material damage somewhere or other in the world. In some regions such events are so frequent and expected that there is some human adjustment to their possible occurrence. Our own country enjoys an equable climate and there is no hint of danger or discomfort in our average climate statistics. But our everyday weather is very variable and we should expect all kinds of unseasonable occurrences and we get them whether we expect them or not. Sometimes it seems that the organization of our lives is keyed to the mythical mean so that the occasional though inevitable extreme catches us woefully unprepared.

Damaging Winds

As can be seen from Table 1 (page 18), wind begins to be disagreeable to a person outdoors when it reaches about Force 5 on the Beaufort Scale. To the farmer, strong winds are more than just a nuisance: they chill and desiccate bare soil, sometimes causing soil erosion in dry areas like the Fens, especially in spring when the ploughed-up fields are still bare (1968 was a bad spring for 'blowing'), they damage crops and endanger animals on exposed hill farms. In more arid lands, dust storms are a common hazard and 'Blowing Dust Reduce Speed' is now seen as a road-sign in parts of Arizona. Calamitous winds, a regular accompaniment of tropical storms, occur more rarely in our latitudes and then only with the deepest depressions.

A wind of gale force is defined statistically as one of Force 8 or more, that is, an average speed of 39 m.p.h. (17·2 m/sec) or more, though this would include gusts probably reaching 50 m.p.h. (22·3 m/sec). On average about 30 gales a year occur on our exposed western coasts but only half this number on more sheltered eastern coasts. The forecasting of gales presents no special problems, being based on the expected development and movement of pressure systems. The Meteorological Office was initially called into being in 1855 for the purpose, still vitally important, of providing gale warnings for shipping.

Few winters pass without a severe gale and reports of uprooted trees, fallen telegraph poles, damaged buildings, dislocated traffic and occasional casualties. Quite unexceptional gales have exposed weaknesses in the design of buildings, causing spectacular 'climatic failures': the gale of 16 February 1962 which de-roofed houses in Sheffield and that of 1 November 1965 which collapsed three cooling towers at Ferrybridge Power Station, Yorkshire, are cases in point. Winds rivalling those of a moderate tropical hurricane have been experienced in this country. In the remarkable Orkneys gale of mid-January 1952, the anemometer recorded 127 m.p.h. (56·7 m/sec) and then blew away. Even inland, though more exceptionally, violent winds of this order have occurred, like the gust of 111 m.p.h. (49·6 m/sec) registered at Cranwell, Lincolnshire, in the severe north-westerly gale (in the rear of a deep occluded low) of 17 December, 1952.

The destructiveness of tropical hurricanes is due only partly to the direct effect of the winds, though these are formidable enough, being capable of lifting houses bodily from their foundations or of sweeping a train off the line (as once happened to a rescue train in Florida). During the New England hurricane of 1938, 275 million trees were destroyed or damaged and when hurricane Hazel passed near New York City in 1954 the Empire State Building set up a rhythmical 16-inch (40 cm) sway (Chapter 7). But the most terrible disasters have been caused by the *storm waves* raised in coastal waters and the flooding of low coastal plains. In 1900, Galveston, Texas, was overwhelmed by 15 feet (4·5 metres) of water and more than 6000 people were drowned. As well as the Gulf Coast – where in recent years, hurricanes Betsy (1965) and Camille (1969) each caused more than $1000 million damage, largely through flooding – the Bay of Bengal is also particularly vulnerable to such visitations.

The more localized havoc wrought by tornadoes is due entirely to their tremendous winds (Chapter 7), which have been known to lift an 80-ton steel railway coach and to drive objects into wood or iron with the force of a bullet. 'Twisters' may cost the United States 200 million dollars and take 200 lives in a year.

Water – Too Much or Too Little

Water in excess or in deficiency characterizes much of the human tragedy in many parts of the world. Even in equable Britain, the climatic record shows that any area is liable to get 100 mm (4 in) of rain in a day at some time or other and equally no part of the country is entirely immune from droughts. Every now and again, serious river flooding occurs due to excessive rainfall, which may take the form of long periods of unremitting rain brought by frequent depressions or of exceptionally heavy downpours of 'cloudburst' type.

Many of our rivers flood pretty regularly during our wetter winters. In the Thames valley above London, many places registered their highest-ever water mark during serious flooding in November 1894, due to persistent depression rain (over 8 in or 203 mm, that is nearly a third of a normal annual total, falling in 26 days). In the exceptionally wet 1960, there were late autumn and early winter floods in Devon: Exeter received 22 inches (about 560 mm) of rain, two-thirds of the mean annual total, in ten weeks. More widespread were the floods of March 1947 (see Frontispiece) which marked the end of that long and bitter winter, when heavy frontal rain coincided with melting snow and frozen impervious ground. The cost to the country was about £12 million. There was something like a repeat performance in March 1963 in very similar circumstances.

Cloudbursts cause more local and sudden flooding. They may occur in lowland country – like the Lincolnshire thunderstorms in May 1920, which unleashed 4 inches (100 mm) of rain in 4 hours and drowned 23 people in the town of Louth – but the effects are always accentuated by the steep slopes and narrow valleys of mountainous regions. Cloudbursts often occur on the lee slopes of high ground, where the up-currents that hold the raindrops in suspension are checked so that the cloud, robbed of this support, spills its content in a great deluge. Such upland flood episodes befell the Dolgellau district of Wales in May 1944 and Bowland Forest in August 1967. Another example of very heavy thundery rain aggravated by orographic lift caused the unprecedented floods of August 1949 in the Tweed valley. Over 5 inches (127 mm) of rain fell in one day on ground already saturated and rivers already swollen after a week of rain. In these Border floods over £1 million worth of damage was reported in Berwickshire alone. The catastrophic Lynmouth flood of 15 August, 1952 was due to a vigorous Atlantic depression with very moist and unstable warm sector air which was forced aloft by the high ground of Exmoor. The phenomenal downpour of 15 August (more than 11 inches or 279 mm in places), rejected by already soaked peaty surfaces, cascaded into the twin rivers East and West Lyn, which descend by steep gradients to join beneath the streets of the little resort. The culvert that normally leads the West Lyn underground into the East Lyn soon became choked with debris and the river took an old course through the town, laying much of it waste, destroying 17 bridges and drowning 28 people.

In another mood, nature sometimes troubles us with floods in flat coastal areas, particularly on our east coasts. Tidal fluctuations depend on the relative positions of sun, moon and earth and, for a given spot, can be accurately forecast for years ahead. But meteorological events occasionally intervene to give water levels much higher than predicted. These *storm surges* are caused mainly by high winds, especially with a long fetch, in a particular configuration of coastline and sea. Our east coast becomes vulnerable when a deep depression has moved across to Scandinavia and northerly or north-westerly gales blow the length of the North Sea. These conditions become dangerous only if they coincide with particularly high tides. Then there may be tremendous piling up of water in the bottle-neck of the southern North Sea, which can be relieved only by flooding, to east and west. On 1 January 1922, sea-level at Southend was 11 feet (3·3 m) above the expected value but this was at low water and no harm was done. But in a similar situation on 7 January, 1928, the water level was nearly 6 feet (1·8 m) above predicted high water and serious flooding affected the Thames estuary and the London area.

The worst storm surge in recent memory was that of 1 February, 1953. A depression had crossed north of Scotland, all the while deepening and giving violent gales (which had already had tragic consequences in the loss of the ferry *Princess Victoria* in the northern Irish Sea). The low then veered south-eastwards, while high pressure built up to the west (Fig. 12.1). With this tight pressure gradient, fierce northerly gales whipped up enormous seas at roughly the time of unusually high spring tides, so that water rose up to 8 feet (2·4 m) above predicted level on our east coast and even more on the Dutch coast. Both to east and west the swollen seas submerged and breached sea walls. In eastern England flooding affected over 300 square miles (nearly 800 km^2), causing nearly 300 deaths, grievous loss of crops and livestock and great damage to property. The Dutch suffered even more, nearly 1800 people drowned, more than 50 000 evacuated: 9 per cent of all agricultural land was inundated.

Statistically a North Sea surge of 1953 dimensions is likely only once in 1000 years but even a lesser combination of these meteorological and oceanographical circumstances would spell danger for low-lying areas like the Thames valley. With river flooding also, it is possible to estimate the frequency of occurrence from past records but the immediate warning depends on adequate short-period weather forecasting

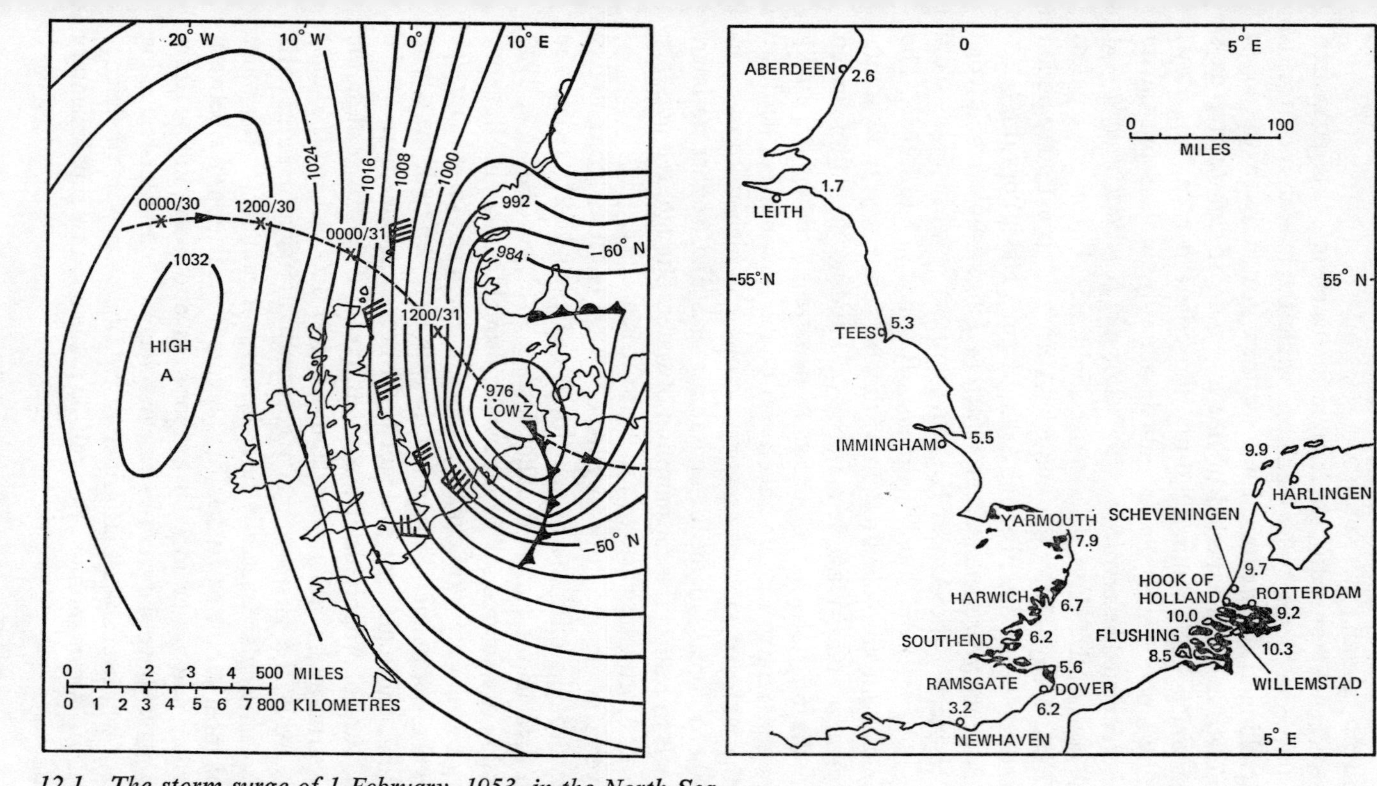

12.1 The storm surge of 1 February, 1953, in the North Sea.
Left: the synoptic situation that morning (the broken line shows the track of the depression with successive positions of the centre).
Right: the extent of the flooding (black): the figures (in feet) represent the surge at high water.

coupled with knowledge of the river and its catchment and the relationship between rainfall and river flow.

At the other extreme, drought is an insidious enemy. It is recognized by the farmer and the water engineer while the man-in-the-street is still marvelling at the continued fine weather. For statistical purposes, an *absolute drought* is a period of not less than 15 consecutive days none of which has 0·2 mm (0·01 in), the criterion of a 'rain day'. If this tiny amount falls, the drought is officially 'broken' but thirsty crops and depleted reservoirs are no better off. A *partial drought* – at least 29 consecutive days with some slight rain but averaging less than 0·2 mm a day – is serious enough. A *dry spell* is a 15 day period without a single 'wet day', i.e. one having 1 mm (0·04 in) of rain.

In this country droughts occur under prolonged anticyclonic conditions (especially the Azores High in summer) or when winds blow persistently from a dry source (e.g. easterly winds in late winter). The likelihood of an absolute drought varies from one a year in south-east England to one in five years in the west and north and one in ten years in the far north of Scotland. A persistent drought eventually becomes apparent in the sun-baked soil of the flower-beds, the dusty roads, the withered look of grass and crops, the high price of vegetables, the outbreaks of heath and forest fires and, sooner or later, the appeals to restrict the consumption of water. Drought is a nuisance here: in less fortunate parts of the world it is a killer.

Dry years in our recent history include 1893 (with a record 60-day absolute drought on parts of the South coast) and 1921 (when the London area had a partial drought lasting 95 days). 1947 and 1959 had notably dry summers. These periodic droughts only underline our rapidly growing water problem. It is less a question of over-all shortage of water than of its mal-distribution. Britain has been built the wrong way round: the high west and north extract an undue share of the available water orographically, so that parts of North Wales, the Lake District and northern Scotland get over 100 inches (2540 mm) annually, while the Thames estuary gets only 20 (508 mm). Large parts of south-east England actually have insufficient rainfall for the water needs of most crops, as studies of the complex relationship between the weather and water loss by evaporation and transpiration from soil and plants have shown. This research has led to a water-budgeting procedure which enables the farmer to calculate the water deficiency during dry summer months: this can be made good by supplementary irrigation.

Hail and Lightning

Although they are the most dramatic features of thunderstorms, hail and lightning affect us relatively little in this country. Most of the hail we experience is small, soft and harmless but once every few years a really vicious hailstorm occurs somewhere in our eastern or midland districts, often with a summer heat thunderstorm. A particularly severe storm struck in Wiltshire on 13 July 1967, with hailstones up to 3 inches (76 mm) diameter, while another notable example was the 'Wokingham' storm of 9 July 1959. Hailstones of even half this size are destructive enough, damaging crops and shattering skylights, windows and glasshouses. Even larger stones have been known to fall in both tropical and continental regions: from India and China there have been reports of 200 or more people being killed in a single hailstorm. Insurance against hail is common in the United States and Germany. Antihail measures are described on page 261.

Lightning activity as such is not usually recorded in detail except in specialist observatories. The direct lightning risk is obviously slight: in the United States, some recent figures of lightning fatalities suggest that the chances of being killed by lightning in a year are about 1 in 1·4 million, while the odds in this country are even less. Most lightning deaths occur in the open, the most dangerous spots during a thunderstorm being on exposed hill-tops, under isolated trees and near wire fences. Indoors the risks have been much reduced by the widespread use of the lightning conductor. Lightning can cause minor injuries, from shocks to burns. It is a fairly common cause of fires, including forest fires and, in this country, it is the most frequent reason for electricity breakdowns, striking overhead wires and giving powerful surges of current which may damage insulation. It is not true that lightning never strikes the same spot twice: tall structures like the Empire State Building and the Eiffel Tower are struck repeatedly.

Frost, Snow and Ice

Frost is a weather hazard we are not likely to escape in any year, though the average number of days with frost varies widely from 10 in our south-western coastal strips to about 70 in parts of East Anglia and 100 over much of Scotland. These figures refer to *air frost* (screen temperature below 0°C or 32°F), which must be distinguished from *ground frost* (grass minimum temperature below 0°C). On frosty nights the coldest air is at the ground: a ground frost alone may spell danger to only the vulnerable tissues of young plants, but an air frost may mean destruction to most plants. Damage to plants – as well as discomfort to human beings – due to low temperatures is always greater if the wind is stronger. Frosts are therefore classified according to the combined effect of temperature and wind, as shown in Table 9, in

Table 9 Classification of Frost

Term	*Screen Temperature °C (°F)*	
	Wind Speed < 10 knots	*Wind Speed > 10 knots*
	(10 knots = 11·5 m.p.h. = 5·1 m/sec)	
Slight	below 0 (32) to −3·5 (26)	below 0 (32) to −0·4 (31)
Moderate	−3·6 (25) to −6·4 (21)	−0·5 (30) to −2·4 (28)
Severe	−6·5 (20) to −11·5 (11)	−2·5 (27) to −5·5 (22)
Very Severe	−11·6 (11) or below	−5·6 (22) or below

which the descriptive terms that figure in the weather forecasts have specific meanings. Yet another distinction is that between 'radiation frosts', which occur at night under clear skies, require calm and depend much on local conditions and the more persistent and widespread 'advection' or 'wind frosts', which occur day and night in very cold (A or Pc) air and are often accompanied by strong winds.

The farmer is little worried about winter frosts, which in fact aid the plough by breaking up the soil. He most fears the late spring frost that catches young plants at a susceptible time after earlier clement weather has encouraged growth. Fruit crops are especially sensitive to late April and May frosts. Frost is a nuisance to others besides farmers and fruit-growers. It is a problem for building work involving the casting of concrete and for the railwaymen concerned with keeping outdoor mechanisms operative and it is a worry to motorists whose vision is impeded by frosted windscreens and rear windows in addition to the danger of frozen radiators. There are answers to these problems; even to those that stem from our traditionally bad plumbing systems which make severe and prolonged wind frosts a headache to householders (though a boon to plumbers).

There are three lines of counter-attack in the battle against destructive frost: it can be forecast, it can be minimized by protective measures or it can be forestalled by prior planning. There are various methods for predicting night minimum temperatures, most of them requiring an estimate of the likely cloudiness and wind. The risk of radiation frost exists in Polar air from autumn to spring, whenever evening skies clear and the wind drops. Frost warnings are given prominence in all official forecasts.

Despite our addiction to Christmas cards portraying snow scenes, snowfall makes up only a small proportion of our total precipitation outside highland regions. It takes 10 or 12 inches of snow to provide the equivalent of an inch of rain. Children have scant opportunity of snowballing in low-lying parts of the south-west, where snow falls on only about 5 days a year and much of this amounts to no more than a few flakes. The chances increase eastwards (10–15 snow days on average in south-east England) and northwards (30–35 in low-lying northernmost Scotland). Snow day incidence increases rapidly with elevation and highland villages may have 40 or 50 a year. Snow rarely lies long in the south but persists for 100 days or more in parts of the Scottish Highlands. Snow is not common at Christmas time, though every now and again the Christmas cards are completely vindicated. It is surprising how many places have most snow days in March or at least as many then as in January or February. The snowstorms of 4 March 1970 (over a foot of snow fell in 10 hours over much of central and south-east England) caused widespread chaos. April too can provide quite a severe snowfall, like that of 25 April, 1950.

A good fall of snow is a delight to the children, but a great nuisance to the community as a whole. It places an extra burden on the municipal authorities who must salt city streets and spread gravel on frozen, compacted snow. Strong winds make matters worse by piling up drifts which may be 7 or 8 m deep (see Plate 21), blocking roads and railway lines (especially in cuttings). Visibility during a heavy snowfall is practically nil and this too adds to the hazards of transport. Isolated villages may be cut off for days and losses of sheep on the hill farms are often grievous. High mountain dwellers live with the danger of *avalanches* (Plate 22). These occur usually when unconsolidated powdery snow accumulates rapidly in great quantities and is displaced by strong winds, or when a rise of temperature half-melts and thus loosens the snow masses. A medium-sized avalanche releases something like 2000 tons (over 2 million kg) of snow in less than a minute: combined with the displacement of air (the 'avalanche wind') the destructive force is terrifying.

In Chapter 3 it was pointed out that many cloud droplets are in the supercooled state. This is a barely stable condition. In the laboratory water can be supercooled, without ice formation, as long as it is not disturbed: if it is agitated, freezing is instantaneous. Similarly in the atmosphere, if supercooled droplets are disturbed, as by a moving aircraft, freezing takes place on it and this is the root of the problem of *aircraft icing*. Ice may form on the wings and windshield or in the engine intake. The general effect is to add weight and so alter the aerodynamic properties of the machine that flight is impeded or, in extreme cases, totally prevented. Icing may be severe in large cumulus and cumulonimbus and thick frontal cloud. An aircraft is also in danger when flying in cold freezing air into which rain is falling from warm front cloud above: this freezes rapidly on to the cold aircraft as glaze.

Containing the Weather

There are ways of containing weather as an enemy and human beings have practised some of them since earliest times with more or less success. The retreat from inclement outdoor weather into the equable shelter of the cave was perhaps the first step in the long search for the optimum indoor environment of the properly insulated and weather-proof dwelling. The old-fashioned cellar performed the function of the modern refrigerator and, in the absence of both, the sensible housewife knows that leaving milk or butter inside an earthenware container kept saturated with water (thus making use of the cooling effect of evaporation) will keep them fresh on summer days.

Not much can be done about the enemy when he attacks on a large scale. In the face of the hurricane, we can only ensure the building of strong structures, the avoidance of floodable land for housing, the provision of emergency food and water supplies and the evacuation of people and livestock from vulnerable coastal areas when warnings are issued. The United States hurricane warning system, helped now by the satellite cameras, is very efficient. Tornadoes however spring up without warning and little can be done except to take cover and hope for the best.

On a smaller scale however the wind can be tamed, for example by the planting of shelter belts and windbreaks on the farm, which provide valuable protection for livestock, crops and buildings, and help to check the blowing of light soils. Cutting down the wind speed means more than reducing mechanical damage to crops, for it also keeps temperatures a little higher and slows down evaporation and transpiration. Shelter belts must not be planted too densely or they defeat their own purpose by setting up eddies with damaging down-draughts on the lee side: solid walls are unsuitable for the same reason.

An effective windbreak is of medium density, allowing air to filter through. Shelter belts are planted at right angles not necessarily to the prevailing wind of the locality but to the damaging wind: this may be easterly or north-easterly in East Anglia or canalized along a valley. Suitably planted and oriented, a shelter belt confers significant protection to a distance of up to 20 times the height of the barrier in the downwind direction and also to a much smaller extent upwind.

Protection against river and coastal flooding is partly an engineering problem. Sea defences can be strengthened: river flow can be regulated, though this may mean storage reservoirs, and a river can be trained to behave itself through endyking or dredging, or the cutting of flood relief channels. There are other obvious precautions: again houses should not be built on floodable land, bridges must be constructed to withstand the river at its highest, the afforestation of steep slopes will slow up run-off. But our water problem needs to be seen as a whole: flood waters run off to waste in highland areas at one time while lowland areas cry out for water at another. There are few effective ways of containing the drought problem. True, it is possible to reduce evaporation from reservoirs by sealing the water surface with monomolecular films of cetyl alcohol, but this is only a palliative measure. More positive thinking in the context of a national water policy seems to be necessary.

Given due warning, the fruit grower can take certain precautions to protect his crop against frost. These measures seek either to prevent heat loss at night or to add heat artificially. The first category includes the use of protective coverings – the gardener's customary method: smoke canopies (produced from oil burners) are less popular now than they were, being too easily shifted by light winds. Even total immersion by flooding with water has been tried but, even where it is feasible, only a few plants take kindly to this treatment. Among methods of adding heat are continuous spraying with water (the latent heat released as this fine spray freezes on the crop surfaces keeps the temperature at 0°C, which is not dangerous) and bringing down warmer air from above through artificial turbulence generated by large electrically driven fans (a costly operation). Direct heating by strategically placed burners is a better proposition where (and if) fuel is cheap. These devices give protection of the order of 1–2°C only. A combination of protective covering by cloches or Dutch Lights and water spraying on to the glass is probably most effective. But the best way to protect fruit is by careful choice of orchard site in the first place: the battle against frost has already been half-lost if the orchard lies in a frost hollow.

Anti-freeze substances like ethylene glycol, which lower the freezing point of a coolant to an extent depending on the concentration, relieve the motorist of anxiety concerning his radiator cooling system and in aerosol spray form will rapidly remove ice from his windscreen. De-icing substances are also part of the answer to aircraft icing, as are methods of circulating engine heat, but pilots are also taught procedures that will take them out of the danger zone as quickly as possible in various icing situations. Sensible householders know that they should insulate water tanks and lag pipes in the loft and ensure that no taps are left dripping in frosty weather since this is the surest way of blocking an outside waste-pipe with ice. Even more desirable are house designs that make these precautions unnecessary: there is, for example, no need to have any water pipes external to the house. Underground heating is already used to maintain certain road bridges and approaches usable in frost and snow conditions and keeps the pitch at more than one football ground playable at all times. Drifting snow remains a hazard on many roads and railway tracks and snow-fences are sometimes used as a counter-measure, though they must be carefully sited or the eddies they themselves create may actually increase the drifting.

Mastery of the Weather?

Containing the weather in limited ways is one thing, mastery of the weather is another, exploitation of atmospheric resources for human ends is something else again. Our activities certainly influence the atmosphere, so far mainly on a local scale, usually in unintended ways and sometimes with unwelcome results: air pollution is the most obvious example of this, though certain other effects (like the urban heat-island) may be regarded as beneficial. Yet there are suggestions that some man-made influences may act globally and over long time periods, thus creating a climatic change. Some authorities believe that a progressive increase in the carbon dioxide content of the troposphere due to increasing use of carbonaceous fuels was responsible (through its enhancement of the Greenhouse Effect of the atmosphere) for an overall warming of the order of 1°F (0·6°C) of surface temperature that began late last century. Unfortunately this temperature trend has reversed itself since about 1940 and, while there is no reason to suppose that CO_2 concentrations are falling, other influences, notably increased dustiness in the atmosphere (due to agricultural and industrial activity?) which would deplete radiation income, may be sufficient to tip the balance the other way. It has also been suggested that aircraft vapour trails at upper tropospheric levels are increasing, perhaps by ten per cent, the cirrus cloud cover in the air lanes near large airports and this could appreciably reduce incoming radiation. The direct warming of the atmosphere as a by-product of our activities may be a factor to be reckoned with, although estimates differ as to how important or imminent this is.

These are inadvertent effects and much uncertainty surrounds them, since the necessary measurements are sometimes hard to make and their representativeness is often in doubt. But to what extent can the weather be consciously controlled? Is it possible, at a time of energy crisis, when irreplaceable fossil fuels are being used up at an ever-increasing rate, to utilize energy at present largely going to waste in the atmosphere? Given our imperfect knowledge, are there dangers in 'tampering' with the atmosphere, because of the possibility of unforeseen repercussions? These are serious questions to which answers are rather urgently required.

Energy from the Atmosphere

Although our planet intercepts only a tiny proportion of the sun's energy output, this endowment is nevertheless extravagantly generous, being equivalent under favourable conditions to the power of a kilowatt electric bar fire per square metre of surface. Small-scale attempts to utilize solar energy date back a long time but it is only recently, as shortages of fossil fuels loom nearer, that serious calls have been made for an effective research and development programme for the tapping of an energy source that is basically free and does not pollute the environment. Despite obvious problems, like the intermittent nature of solar output and the need for some kind of topping-up system, there seems little doubt of the feasibility of solar energy as a resource, particularly in developing countries where, it so happens, both the need and the opportunities for meeting it are greatest.

Primitive utilization of solar energy has taken forms like the drying of crops and the evaporation of sea water for salt extraction (still practised as far north as the south coast of Brittany). More modern applications convert solar into thermal energy by the use of 'collectors', which are either of the flat-plate or the focussing type. The latter are basically parabolic reflectors and their use is confined practically to small solar heaters or cookers, able to boil water in 20 or 30 minutes, which are commercially available in various tropical countries and, on a larger scale, to experimental solar furnaces like that at Mont-Louis in the French Pyrenees, where movable 'sun-tracking' concave mirrors create a temperature of 3000°C at the focal point. Flat-plate collectors are used in the dozen or so existing 'solar houses', taking the form of a black surface under glass plates on the roof (which is usually pitched equatorwards): water pipes embedded beneath the black receptor extract the heat and carry it for storage either into a water tank or to a container filled with crushed rock (as used in electric night storage heaters), from which circulation throughout the house is effected.

In many areas, the need is less for solar heating than for solar cooling, especially over much of the United States where large amounts of electricity are consumed in air-conditioning plants to keep indoor temperatures down to tolerable levels in summer. Solar space-cooling systems and refrigerator systems for food already exist in the experimental stage. Also promising is the solar distillation plant (small commercial units are already available) for the desalination of sea water. Other applications lie in the conversion of solar energy to forms other than thermal: the production of electric power (solar batteries) is already commonplace in space vehicles but on the earth's surface would be too expensive for large-scale use though feasible for specialized low-power purposes. Solar-powered pumps would be useful for irrigation water or the watering of stock. Sun power could also be used to activate certain chemical reactions (e.g. the melting of sodium sulphate in order to utilize the cooling effect of fusion) or bacteriological systems (to produce certain liquids or gases).

It is a long time since man first learned to harness the power of the wind on the small scale of the expanse of cloth that propels a sailing-vessel or the circle swept by the arms of a windmill. The wind is ever-present and free and its further exploitation would seem an attractive idea but in fact it is an irregular and unreliable entity and it is not possible to store its energy directly. The windmill is now a picturesque landscape feature of historical interest and sailing-vessels are a sign of necessity in less developed areas or of leisure opportunities in more affluent parts of the world. But given the right wind regimes (the blowy westerlies or the trade wind belt) and the most suitable sites within them (open plains on windward coasts or exposed hills or plateaus), wind-driven generators supplying a few hundred kilowatts are feasible.

The scale is still small and wind power is easily out-matched by other sources where these are available. But in the most suitable locations and if there is some back-up provision when the wind strength falls below the critical level for the appliance, wind generators have some future and indeed may be found in operation today in parts of the Netherlands,

Denmark, Greece, Israel and Australia. The output can only be used immediately, unless research results in a suitable storage battery.

Unlike the somewhat random power of the wind, the flow of water organizes its energy of motion conveniently into channels. From Saxon times this power has been used to drive the mill wheels for the grinding of corn and in mediaeval times the rivers worked the tilt hammers for crushing rock and extracting ore and operated the forge bellows for the iron industry. Since late last century, from modest beginnings in the French Alps, we have learned to convert water power into the more convenient and transportable form of hydro-electricity. Installations vary in size from small plants in mountain streams whose turbines will light a village or serve minor industries to large-scale works involving an enormous fall of water, which may necessitate barrages and the impounding of storage reservoirs.

As far as Britain is concerned, hydro-electricity is of small importance in England and Wales but meets about a third of present power demands in Scotland, where there is further capacity as yet untapped. Our river regimes are favourable in that flow is greatest in winter when the demand for electricity is also at its greatest (unlike more Alpine areas where winter, with water locked up in snow and ice, is the time of low water). There is, of course, a limit to the extraction of water power in economic terms: France is already nearing the maximum exploitation of its hydro-electric resources.

Enough Water for All

A man will not survive without water for much longer than a week. Early man had to live by the banks of streams to satisfy his daily needs for his family and his animals. Mediaeval woollen industries needed the water for scouring, washing and dyeing as well as to provide power for the mill wheels. Today 'civilized man' in developed countries is accustomed to water on tap and may forget his ultimate dependence for this cheap commodity on rain and rivers and underground supplies. Apart from growing domestic consumption, the thirst for water on the part of agriculture, industry, electricity generation, inland navigation, even certain forms of recreation, imposes an increasing strain on our resources.

At first sight these seem adequate, even generous. The United Kingdom as a whole has a mean rainfall (1079 mm or 42·5 in) rather higher than for the earth as a whole. For England and Wales, taking into account the average amount of water present in the ground and in water bodies and the average evaporation loss, it seems that the available supply exceeds the present demand nine or ten-fold and there are, theoretically, some 900 gallons (4000 litres) of water per day per head of the population. This statement conceals the problem that water availability varies greatly from year to year and from region to region within the country. According to some estimates, demand by the end of the century will not be so far short of the available yield. And in some areas present rates of water abstraction already exceed what is called the usable dry weather flow of the river and the situation is saved only by the successive re-cycling of water.

On average in this country we each use domestically some 35 to 40 gallons (160 to 184 litres) of water a day. In some areas the industrial use of water, as a raw material in chemicals or paper, or for feeding boilers or for cleansing or cooling purposes, may far exceed this rate of consumption per head of local population. Cooling water for thermal power stations is easily the largest single item in the list of licensed abstractions in England and Wales, though the use of re-cycling techniques in cooling towers is becoming more common. Nuclear power stations are nearly all situated on coasts because of their huge water demands. Agriculture requires water increasingly for irrigation as well as for watering livestock. A certain minimum river flow is necessary to keep canals viable and to prevent the silting-up of river estuaries. The water problem is largely one of increasing consumption rather than of diminishing supplies, although there are anxieties about the over-exploitation of ground water reserves so that water levels in some important aquifers (water-rich rock strata) are seriously depressed. This has happened in the London Basin and also in some coastal areas to an extent that has allowed the entry of saline water.

The problems are not insoluble. The metering of domestically used water might well check extravagance: the major non-domestic consumers are already metered. The further extension of recycling methods would ease the situation in many areas. The regulation of river flow by reservoirs (which might also provide hydro-electricity and water sports facilities) high up in the catchment areas and equally the control of ground-water is obviously desirable: over-pumped aquifers will replenish themselves naturally from rainfall but artificial re-charging is also feasible by leading surplus river water into the very boreholes sunk initially for water abstraction. Barrages across bays and estuaries (Morecombe Bay, the Dee estuary, the Wash) which will create fresh-water lakes are being studied but the issues are complex and far-reaching. Desalination of sea water on a large scale is an expensive proposition at present. Whatever is done, a nationally conceived solution for what are basically regional problems would seem inescapable, for, as has already been intimated, flood control in the uplands could mean drought relief in the lowlands. Because of the physical build and economic development of this country, the major regions of supply and demand do not coincide. A considerable movement of water across natural (and administrative) boundaries, by canal or pipeline, would be an essential part of any such plan.

The implications are far-reaching and complex. Any satisfactory answer to the water problem must take into account not only consumer demands and the maintenance of high water quality but also the requirements of conservation and general amenity. In 1970, 315 water undertakings could be counted within the United Kingdom, but the effective handling of our water resources calls for a smaller number of larger bodies. The notion of 10 multipurpose regional water authorities, responsible for supply, quality, conservation, navigation, sewage, amenity, etc, is part of a Government plan which should come into operation in 1974, to coincide with the reform of local government areas.

Cloud Seeding

In many parts of the world where water is in naturally short supply the prospect of artificially increasing the rainfall seems to offer the greatest challenge to the would-be weather controllers. There is nothing intrinsically new in the idea of rain-making for in primitive societies this was a routine task performed by the resident magician. But modern rain-makers appeal not to mystical but to meteorological principles, which exist in the precipitation mechanisms described in Chapter 3. Scientific experiments began in the United States in 1946 ('Project Cirrus') under Langmuir and Schaefer and soon afterwards in Australia under Bowen. Work has continued in those countries and many others, including England, but the main effort has been in regions of marginal rainfall.

Rain cannot, of course, be conjured out of a cloudless sky, nor will every cloud be induced to part with its moisture. We have seen that in temperate and high latitudes the Bergeron process will operate only when ice particles and supercooled water droplets co-exist in the same part of the cloud. A cloud extending to these levels may well rain of its own accord: a cloud not reaching freezing level will not rain at all. But a cloud that pushes above freezing level (i.e. contains supercooled droplets) but not high enough for the spontaneous formation of ice crystals has possibilities, provided it can be *seeded* with freezing nuclei of some kind.

It was on the basis of this reasoning that Langmuir and Schaefer dropped pellets of solid carbon dioxide ('dry ice') into suitable cumulus clouds from aircraft. This very cold substance cools part of the cloud to below −40°C so that many ice crystals form spontaneously, thus setting the stage for the Bergeron mechanism. Soon afterwards, Vonnegut seeded clouds with silver iodide which, having a crystal structure much like ice, is a more effective freezing nucleus which induces freezing at temperatures as high as −5°C (23°F). He showed that it was most economical to generate the chemical on the ground in the form of a smoke (by burning fuel impregnated with a silver iodide solution) and to rely on natural up-currents to carry it into the cloud.

In Australia in 1952 Bowen tackled the rather different problem of 'warm clouds' (Chapter 3) by spraying them with water, on the principle that this would introduce the larger droplets necessary for the operation of the coalescence mechanism. In Pakistan, salt solution was sprayed into clouds to provide the large condensation nuclei that would generate the larger droplets. A related method was tried in East Africa (in order to maintain reservoir levels) where naval rockets containing salt particles were fired to burst and scatter their contents at suitable heights within the cloud.

From the earliest experiments of Project Cirrus it was clear that in many cases individual seeded clouds grew rapidly and precipitated while untreated clouds remained unchanged. These apparent initial successes led to a great surge of rain-making activities, including those by commercial firms in the United States. Claims were made that it was feasible to augment natural rainfall by 20, even 30 or more per cent. But today, in the third decade of effort, the experts are far more cautious in their pronouncements for some seeding projects have given inconclusive results, while others seem to have been followed by a decrease in rainfall. Clearly it is difficult to evaluate the success of a seeding operation over a large area: there are uncertainties as to whether the seeded clouds might have grown sufficiently to precipitate naturally or even whether silver iodide released from ground generators (and this is the method mainly used) actually reaches the clouds in suitable condition and concentration. After the waning of early enthusiasms, opinion seems to be that an increase above expected rainfall of the order of 10 per cent is likely but experimental work goes on, such would be the benefits of even a modest bonus of additional rainfall.

Strangely at first sight, rain-making techniques are also used for the prevention of another type of precipitation – hail. Seeding of a potential hail cloud with silver iodide is intended to produce a large number of very small hailstones which will inhibit the growth of very large ones. There is particular incentive to try hail suppression techniques in vineyard areas and success has been claimed in the Beaujolais in France for the method of firing the silver iodide into the cloud by explosive rockets. A substantial reduction in hail damage through similar techniques has been reported from the Caucasus region of Russia. 'Hail-shooting' by cannon (and bell-ringing before that) is an old practice, now defunct, but the idea that explosions may shatter large hailstones persists in the use of rockets (without silver iodide) in Italy and elsewhere, with what are said to be worthwhile results.

Artificial nucleation by silver iodide has also been tried as a means of dissipating supercooled fogs and works effectively, much of the moisture falling out as a minor flurry of snow. The more common 'warm fogs' however can be eliminated only by evaporation and this requires fuel. Oil burners were used on certain British airfields during the Second World War but the method was costly. There is also the notion that seeding parts of a hurricane would, by ensuring latent heat release over a broad area, reduce the pressure gradient and thereby the wind speed: results of such experiments so far are inconclusive.

Whose Atmosphere?

There is general agreement that weather affects our lives in many ways and that we would be a lot better off if we understood it more completely and especially if that understanding led to reliable longer-range forecasting. Even at the present time, the weather services justify the money spent on them many times over, in terms of savings to agriculture, industry, power supply, etc and, while it is difficult to assess these matters, a benefit-cost ratio of 20:1 has been suggested for the United Kingdom and may be regarded as applicable to most national meteorological services. Moreover in the near future when the benefits of World Weather Watch and GARP begin to accrue, the ratio should become even more favourable. But about the possible benefits of attempts at conscious weather modifications, even if they are technically feasible, there is much less confidence and little unanimity.

Even in the case of 'rain-making', to which considerable effort has been applied and for which some measure of success has been claimed, there is reason to pause and reflect. If at any one time all the moisture contained in the atmosphere were to fall out and remain on the earth's surface it would result in a layer of water on average about one inch deep. The amount of precipitable atmospheric moisture is limited and, if more rain can be induced to fall in one place, less will be available elsewhere. What is good for some farmers may therefore be bad for others and, even within the same region, increased rainfall may benefit some consumers and disadvantage others. Obviously the atmosphere dispenses both benefits and disadvantages but human beings may evaluate these very differently. This raises the awkward question of whether atmospheric resources are common property, even assuming that they can be exploited. This is not always a problem: we can all bask in the same summer sunshine (provided there is room for all of us on the beach). But on the other hand there are conflicting demands on the limited water supplies that derive from precipitation.

In an environment-conscious age, we are justifiably beginning to query the right of an individual or organization to pollute someone else's atmosphere. Any non-smoker who has sat in a cinema or theatre and suffered his neighbour's cigarette smoke curling incessantly into his nostrils will appreciate this point. Anyone unfortunate enough to live on a busy road willy-nilly breathes pollution from someone else's car exhaust. The farmer must put up with pollution made in the nearby town. Because of the accident of location within the global circulation, one country may receive part of the pollution of another.

There have been many recriminations, local, regional, international, about pollution, which is at least an inadvertent evil. But when it comes to purposeful modification of the atmosphere, the social, legal, even political implications require a good deal of thought. Who is to be allowed to 'tamper' with the atmosphere? Can this be left to private enterprise or is it a matter for governments or even supra-governmental agencies to decide? In the United States, commercial cloud-seeding ventures have been commissioned by farmers who need the rain but legally opposed by holiday resort interests which are better off without it. There have been court-room battles to prosecute rain-makers because cloudbursts or hailstones have adversely affected property or crops. Decisions in these legal cases have gone either way, not least because of the difficulty of deciding, in the present state of the art, whether the attempted modification did indeed give the results complained of. Again perhaps it is not too fanciful to see the implications extending into the unpleasant possibility of weather modification used as a weapon of war.

Another aspect of this problem is our ignorance of the consequences of large-scale experimentation with the atmosphere. From time to time there is talk of removing snow and ice by coating the surfaces with soot or coal dust (thereby reducing the albedo and encouraging the absorption of solar radiation). The melting of the Greenland ice cap would raise world sea level about 7 m (23 ft) thereby altering many familiar coastlines. To take a less extreme suggestion it would be not at all far-fetched to use this method to keep the central Arctic ice-free, since it is estimated that only a relatively small energy input would suffice to melt the ice floes. It is uncertain if the ice would re-form or how frequently the treatment would have to be applied. The result would be much higher winter temperatures (around freezing point nevertheless) and at least a weakening of the north polar anticyclone and of the pressure gradient across middle latitudes. The consequences might entail substantial shifts of depression tracks and the distribution of rainfall. Similar fears were expressed following a Russian proposal to irrigate large areas around the Aral and Caspian Seas with water diverted from the north-flowing rivers Yenisei, Ob and Pechora: the loss of this fresh water would increase the salinity of Arctic waters which would lower the freezing point and reduce the ice cover. In fact, only a small porportion of the river water would apparently be diverted and the consequences are not likely to be significant. Some other suggestions seem insecurely founded, for example the creation of a Saharan lake with water from the Congo: despite high evaporation, the stability imposed by the pressure regime would inhibit rainfall nearly all the time.

A saving grace in most of such proposals as removing ice caps and diverting warm ocean currents is that the techniques are very uncertain and enormously costly, which is a good reason for surmising that nothing will be done. Certainly, as most responsible scientists argue, nothing should be done, at least until the fullest possible understanding of what may be involved has been reached through careful and thorough investigation. 'Knowledge is power', it is said, but knowledge should also lead to prudence. The energy that man can deploy is still very puny compared to that available to nature. To talk of controlling the weather is vanity, as far as can be seen at present, and to tinker with it would be difficult and maybe dangerous. Partnership (cautiously approached) is another matter. To make the most of the free gifts the atmosphere confers, to protect ourselves effectively from its inclemency and perils and to order our affairs in the light of improved fore-knowledge, is perhaps as much as we can expect. But it is a great deal nevertheless.

Suggestions for Further Reading

On Observing

Meteorological Office, Met. O. 805, *The Observer's Handbook*. H.M.S.O., third edition (1969).

Meteorological Office, Met. O. 670, *Instructions for making the Stevenson type of Thermometer Screen*. H.M.S.O. (pamphlet).

Royal Meteorological Society, *Running a School Weather Station* (pamphlet, available from the Society: for address, see under Journals, below).

On Organization

Meteorological Office and Central Office of Information, *Your Weather Service*. H.M.S.O., third edition (1959) (booklet).

Meteorological Office, Met. O. 515, *Instructions for the Preparation of Weather Maps*. H.M.S.O., fourth edition (1967) (booklet).

On Weather in General

D. H. McIntosh and A. S. Thom, *Essentials of Meteorology*. Wykeham, 1969 (a more technical introduction, for which some Sixth Form physics and mathematics are an advantage).

D. E. Pedgley, *A Course in Elementary Meteorology*. H.M.S.O., third impression (1968) (written primarily for weather observers but useful for anyone).

C. E. Wallington, *Your Own Weather Map* (pamphlet, obtainable from Royal Meteorological Society: see under Journals, below).

On Weather Lore

R. Inwards, *Weather Lore*. Fourth edition, Rider 1950.

John Claridge (Shepherd), *The Shepherd of Banbury's rules to judge of the changes of the weather, etc*. Annotated by G. H. T. Kimble. (Reading University, 1941).

On Climate

A. Austin Miller, *Climatology*. Eighth edition, Methuen, 1957.

G. Manley, *Climate and the British Scene*. Collins, 1952.
H. H. Lamb, *The English Climate*. E.U.P., 1964.

On Weather and the Community

W. J. Maunder, *The Value of the Weather*. Methuen, 1970.

Journals

Royal Meteorological Society, *Weather* (monthly, from Cromwell House, High Street, Bracknell, Berks RG 12 1DP).
National Society for Clean Air, *Clean Air* (quarterly, from 136 North Street, Brighton BN1 1RG).

Index